The Candle Revisited

The Candle Revisited

Essays on Science and Technology

Edited by
P. DAY
and
C. R. A. CATLOW

Oxford   New York   Tokyo
OXFORD UNIVERSITY PRESS
1994

Oxford University Press, Walton Street, Oxford OX2 6DP

Oxford   New York   Toronto
Delhi   Bombay   Calcutta   Madras   Karachi
Kuala Lumpur   Singapore   Hong Kong   Tokyo
Nairobi   Dar es Salaam   Cape Town
Melbourne   Auckland   Madrid
and associated companies in
Berlin   Ibadan

Oxford is a trade mark of Oxford University Press

Published in the United States
by Oxford University Press Inc., New York

A catalogue record for this book is available from the British Library

Library of Congress Cataloging in Publication Data
The candle revisited: essays on science and technology/edited by P.
Day and C.R.A. Catlow.—1st ed.
1. Science   I. Day, P.   II. Catlow, C. R. A. (Charles Richard
Arthur), 1947–   .
Q158.5.C35   1994   500—dc20   94–2449
ISBN 0 19 855835 X

Typeset by Footnote Graphics, Warminster, Wilts
Printed in Great Britain by
Biddles Ltd, Guildford and King's Lynn

PREFACE

Since they were established by Michael Faraday in 1826, the Royal Institution's programme of Friday Evening Discourses has enabled many famous names in science to describe their work to a wide audience consisting of scientists and non-scientists alike. Traditionally, such Discourses are illustrated profusely by pictures, exhibits, and especially by demonstrations. Since the nineteenth century, articles based on the Discourses have appeared in the *Proceedings of the Royal Institution*, published annually. Now, in association with Oxford University Press, the Royal Institution is making a selection of these articles available each year to a wider public in an attractive soft cover format. Like the Discourses themselves, these articles cover some of the most fascinating aspects of contemporary science in a lively and approachable way designed for the general reader. Wider appreciation by the public at large of the importance of science, engineering, and technology is generally agreed as vital to an advanced society. In bringing this selection of authoritative but readable accounts to a larger audience, the Royal Institution is playing the role in this endeavour that has always been part of its programme and its ethos.

London

February 1994

P.D.

C.R.A.C.

CONTENTS

PLATES

Plates fall between pages 38 and 39, 54 and 55, 64 and 65, and 86 and 87.

CONTRIBUTORS

P. W. Atkins
University Lecturer in Physical Chemistry and Fellow of Lincoln College, Oxford.

John V. Bartlett
Mott MacDonald 20–26 Wellesley Road, Croydon.

R. R. Chianelli
Materials Research Society, Exxon Research & Engineering Company, Annandale, New Jersey, USA.

Peter Day
Director and Resident Professor of Chemistry, The Royal Institution, London.

H. J. Evans
Medical Research Council, Human Genetics Unit, Western General Hospital, Edinburgh.

Robert Greenler
Department of Physics, University of Wisconsin, Milwaukee, Wisconsin.

Paul Murdin
Royal Greenwich Observatory, Madingley Road, Cambridge.

R. H. Williams
Department of Physics and Astronomy, University of Wales, Cardiff.

The candle revisited

P. W. ATKINS

There is no better, there is no more open door by which you can enter into the study of natural philosophy than by considering the phenomena of a candle

So Michael Faraday began his classic series of lectures at the Royal Institution a century and a half ago. However, although Faraday had an unparalleled sense of matter, and an unequalled facility for sharing his extraordinary sensitivity towards matter with the eager and the curious, his inner eye was blinkered by his time. His was the milieu of classical physics, where even the concept of an atom was not universally accepted, where the discovery of the electron lay decades in the future, and where there was no inkling of the weirdness of quantum theory. In this single article, I shall wander across Faraday's ground with a later eye, and share what he could not see despite the awesome clarity of his vision.

Faraday's approach was to present and comment on phenomena. With our later knowledge of the inner workings of the world, the modern paradigm of scientific exposition has shifted towards providing comprehension by forging links between the imagined and the perceived. A candle provides a vehicle for exploring that link. It also acts as a vehicle for crossing another bridge, from the organic to the inorganic. A candle is an everyday Medusa. When I was recently in Kazan, in the Tartar Republic, there was great puzzlement about why the stony face of Medusa should appear over a certain chemist's door. But all chemists are Medusas. A candle is a Medusa, for the fate of its burning wax, an organic material, may be to enter the inorganic world as limestone once its carbon atoms have been temporarily liberated by the flame. Nevertheless a candle flame is not an unremitting Medusa, for the liberation of the flame allows a dormant carbon atom to return to life.

Even before a candle is lit, there is a puzzle worth noting, and the recognition of a puzzle is often a clue to the direction in which deep understanding can be found. The particular puzzle in this instance is why solid candle wax is white while molten wax is as transparent as water.

The opacity of solid wax is a result, we now know, of the shapes of wax molecules, and in particular of their ability to lie together in a solid to form local regions of ordered matter. These organized regions are effectively tiny crystals, which scatter the light that falls on them. When wax melts, these crystalline regions disappear and the scattering no longer occurs. Exactly the same phenomenon accounts for the whiteness of snow despite the transparency of water. Indeed, although the carbon dioxide produced by a candle is colourless, solid carbon dioxide is also white when it is finely divided.

When a candle is lit, we enter a much richer domain of enquiry. Indeed, the flame encapsulates several regions of modern scientific discovery (Fig. 1). Thus, the region of the incandescent flame is the region which epitomizes the content of quantum theory, a twentieth-century paradigm of science; perhaps *the* paradigm of science for this century. Nearby in the structure of the flame we have the region of turbulence. Here is a major region of modern research, which has become open to investigation only with the availability of powerful computers. Just above the flame is the region where soot is prevalent. This region symbolizes the region of chemistry that has only very recently come into being. Here, there may be (although no one has yet discovered the molecules there), the extraordi-

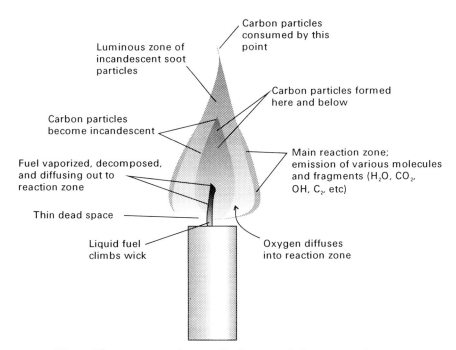

Fig. 1 The structure of a candle flame and the principal processes occurring in each region.

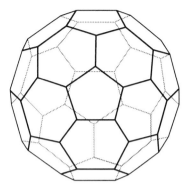

Fig. 2 Buckminsterfullerene, C_{60}.

narily symmetrical molecules of Buckminsterfullerene (C_{60}), the three-dimensional analogue of benzene (Fig. 2). It is very fitting that benzene (C_6H_6), which was discovered by Faraday, should have its higher dimensional hydrogen-free analogue present in more intense and sophisticated versions of regions of Faraday's candle flame. At the foot of the flame is the region of colour, where molecules still have their individual identities, and the radiation they emit is characteristic of their presence. Here, in this cooler region of the flame, energetically excited versions of H_2O, CO_2, OH, C_2, and other fragments of the waxy fuel throw off their excess energy as electromagnetic radiation, and we see their pale blue glow. The same colours of a candle flame also reach far out into the cosmos, for we see similar colours in a comet's tail. Here the head glows with the radiation from excited C_2 and CN fragments, and the tail glows with radiation from excited CO^+ molecular ions. Whereas the energy of excitation of a candle flame is provided by the burning fuel, for a comet the source of excitation is the nuclear candle in the sky, the Sun, and the naked ultraviolet radiation it emits.

The reaction

The candle is also a microcosm of the macrocosm, for its flame is found on a larger scale, as in the conflagration of a forest. Here, as in the flame itself, three kinds of question arise which Faraday seemingly did not think to ask, in public at least. First, what is the driving power of a reaction? In other words, why does this conflagration proceed without needing to be driven? Second, why does the reaction need to be ignited? Why do not all flammable materials burst into flame as soon as they are formed? Third, how does the reaction take place? That is, what is its detailed molecular mechanism? Once we have a mechanism for a reaction, we may be able to study it elsewhere and so reduce the apparent complexity of the world. The conflagration of a forest, like the combustion of a candle flame, is in fact an example of a common type of reaction known as a *redox reaction* (that is, a reduction–oxidation reaction), in which electrons are transferred from one species to another.

Redox reactions are also found in other disguises, as in the corrosion of steel, like that that now encrusts the prow of the *Titanic* as she lies in the deeps of the north Atlantic. However, do not be misled by the superficial differences between the roar of a forest fire and the quiet corrosion in the deep abyss. The two seemingly different reactions are *essentially* the same: they seem different, but that is a reflection of the different personalities of carbon and iron. For instance, the reaction products of water, oxygen, and iron are insoluble solids that neither blow away nor dissolve. Thus, corrosion is much slower than combustion because solids are much less mobile than gases and liquids. Moreover, corroding iron does not glow like burning carbon because the energy leaks away as quickly as the reaction proceeds, and none is available for the relatively slow process of discarding as radiation. Moreover, the reaction is plainly not extinguished by water! The reason for this difference lies in the details of the reaction mechanism by which steel gives way to corrosion.

Why reactions occur

At this point I shall consider the first of the questions that Faraday did not ask: *why do chemical reactions occur?* This is the domain of that great liberator of the human spirit, the Second Law of Thermodynamics. There are in fact three, and only three, primitive springs of change, and each of them is of extraordinary simplicity. Any one, or more commonly combinations of the three, can drive the world forward. All change is a manifestation of one of these inner springs of nature. In short, progress is collapse into chaos.

There are three contributions to chaos:

1. There is the *dispersal of matter* as local accumulations of matter spread. Such a process seemingly drives a gas of randomly moving atoms to expand to fill whatever container it occupies. The reverse of such an expansion has never been observed.

2. There is the *dispersal of energy*, as atoms jostle their neighbours and local accumulations of energy spread. This type of process accounts for the cooling of a hot

object to the temperature of its surroundings. We never observe energy accumulating chaotically in an object that is hotter than its surroundings.

3. There is the *loss of coherence* of motion, as organized atomic motion decays into the disorder of thermal motion. This type of dissipation accounts for the loss of motion of a bouncing ball, for the motion of its atoms becomes a little more disorganized on every bounce, and its motion (the orderly motion of its atoms) decays into chaotic thermal motion. A ball resting on a warm surface has never been observed to start bouncing, even though (because it is hot enough) it possesses enough energy. Atomic motion cannot become orderly spontaneously. If ever a ball were spontaneously to leap into the air, then modern science would collapse. Or at least, the event is so improbable that we can be sure that it will never occur.

So far, we have dealt only with primitive physical events, whereas our concern, like Faraday's, is with chemistry. Chemistry, though, is emergent physics, and had Faraday lived for another 50 years, then he would have understood why his candle burned. And he would have understood in terms of these three primitive physical events. A burning candle, for example, disperses matter as the linked atoms in the compact, relatively orderly chains of carbon atoms in the fuel are liberated as numerous independent CO_2 molecules. Moreover, energy also flows chaotically into the surroundings as it is liberated in the flame. The same process drives our vehicles forward, when the small hydrocarbon molecules used as fuel for internal combustion engines scatter as fragments and release energy.

Chemistry, though, is more subtle than physics, and it is essential to be circumspect. Thus, although a reaction might absorb energy and hence rise to a state of higher energy, the reaction might give rise to a highly disorganized product, and thereby generate net disorder. Thus, in chemistry, some reactions can run in an unexpected direction. Indeed, all the events in this wonderful, glorious world, even those that seemingly generate structure and form, are at root driven by a purposeless collapse into chaos.

One important consequence of the power of collapse into chaos is that, given a suitable link, one reaction can drive another reaction in an unnatural direction. The only criterion is that the increase in disorder arising from the first reaction is greater than the loss of disorder arising from the second reaction. The analogy (and it is only an analogy, as the principles involved are different) is the motion of two coupled weights. If we saw one weight hanging in front of a screen rise apparently spontaneously into the air, then we would infer that it was coupled to a heavier weight behind the screen, and that as the second weight fell it raised the lighter weight. Understanding chemistry, and certainly biochemistry, is essentially seeking heavy weights behind subtly hiding screens.

One example of a 'heavy weight' reaction was very important to Faraday and figured in his lectures. He built a Voltaic pile by stacking together a series of disks of copper and zinc separated by moist paper. On account of the electronic structures of these two metals and of their ions, there is a natural tendency for electrons to leave the zinc metal and to attach to copper ions, so forming copper metal. If the electrons released by the zinc are not allowed simply to migrate from zinc to metal but are taken from the zinc through an external circuit, then they can be used to do work and carry out an unnatural reaction, such as the decomposition of water into its elements.

We have our own internal batteries, our own 'heavy weight' reactions inside us, reactions that drive other reactions in unnatural directions and stave off equilibrium (for equilibrium is death). One very important 'heavy weight' reaction is the loss of a phosphate group from the molecule adenosine triphosphate (ATP) to form adenosine diphosphate (ADP). The chaos created in this step is coupled (by enzymes) to other reactions. For example, it takes about four ATP $\rightarrow$ ADP events to build each link in a polypeptide (a protein); it takes countless but a finite number of such events to form an opinion.

How does the battery get charged? By eating. A sandwich, and the slow combustion-like reaction it undergoes in our bodies, is an even heavier weight; it produces even more disorder in the world, and hence, suitably coupled, can restore ADP to ATP even though that is an order-creating step. And where does a sandwich receive its charge? It is charged by the Sun, the most 'heavy weight' reaction of all for us on Earth.

The initiation of reaction

I shall now turn briefly to another problem that Faraday seemed not to consider: why a candle needs to be ignited. We have seen that a combustion reaction has a natural tendency to occur, largely because it develops so much disorder; so why does it need to be initiated? Why do candles not just burst into flame? Faraday knew that he had to initiate change, but appears not to have known why.

In brief, there is an energy barrier between reactants and products, and only if the reactants can be levered over this barrier will the reaction take place. Thus, bonds have to be stretched and broken before reaction can ensue, and reactants have to crash together. Once again, it is the dispersal of energy that, as well as driving the reaction forward, now holds the reaction back. Chaos is both the carrot and the cart. The reaction can take place only if enough energy accumulates where it is needed: it must jostle together to form local hotspots. Such local accumulations of energy form only rarely, so reactions rarely occur. To cause them to occur more frequently, we have to raise the temperature locally, such as by bringing up another flame to the reaction we want to initiate.

This description of the need for initiation adds to our understanding of the ways in which a candle can be extinguished:

1. We can snuff it out. That, more formally, is the prevention of one of the reactants (oxygen) reaching a second reactant (the hydrocarbon fuel).

2. We can blow it out. The air stream removes hot products and prevents them from supplying local hotspots of energy to help lever the incoming reactants over the energy barrier.

3. We can pour on water. In this action, the bulk water is ripped apart into a gas of molecules, and that energy-intensive process lowers the temperature locally and hence eliminates local hotspots.

How reactions occur

Now I shall consider the depth to which reactions are currently understood. Here we make the transition into the world of explanation that was totally beyond Faraday's comprehension, for the appropriate physics lay decades in the future.

We know that molecules exist, and we can determine their detailed structures in terms of atoms. We also know that atoms exist, because we can see them (or, at least, form compelling images of them, Fig. 3). Of both these entities Faraday had an inkling. However, he did not know of electrons, and in particular he did not know that electrons had the properties of waves. It is of crucial importance for understanding modern chemistry that we take into account this wavelike property of electrons. Specifically, we need to appreciate that the distribution of an electron has regions that differ in sign. If we think of a wave as having regions of positive displacement (which I shall refer to as 'red' regions) and regions of negative displacement ('green' regions), then in a sense classical physics was colour-blind. Yet appreciating the existence of these regions of different 'colour' of the distribution of electrons is essential to our current understanding of the structures of molecules and the changes they

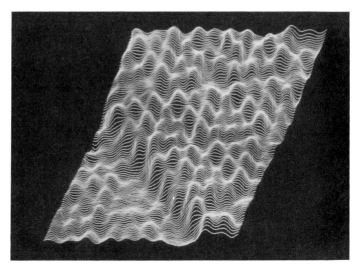

Fig. 3 The type of image that can be obtained with scanning-tunelling microscopy. The sample is of a silicon surface and the cliff is one atom high.

undergo. It is a relatively simple matter to know that an electron is distributed like a cloud; it is deeper and essential to know that this distribution is 'coloured'. Thus, molecules like the carbonate ion shown in Fig. 4 (one of the possible fates of carbon from a burning candle) are described by distributions where colour is essential to understanding. One consequence of the existence of colour is that we can anticipate that symmetry considerations play an important and subtle role.

The understanding of symmetry, and its mathematical description in terms of group theory, has been one of the most profound developments in physics this century. Its role in the analysis of elementary particles is well known (and abstruse). Faraday was aware of symmetry, just as we all are intuitively; but it became powerful when it was formalized, when numbers could be attached to intuitive notions. Symmetry considerations are central to our understanding of the structures of solids, of the structures of molecules, and more recently to the reactions that molecules undergo. The reactions in the candle flame can only be fully understood in terms of the symmetries of species and of their electron distributions.

I shall now turn our attention to the steps by which a chemical reaction occurs. My aim here is not to go into detail, but merely to give an impression of the detail in which we understand the processes by which reactions occur. Faraday knew that reactions occurred; but he had no understanding of their inner mechanism.

Perhaps the simplest type of reaction is the combination of radicals,

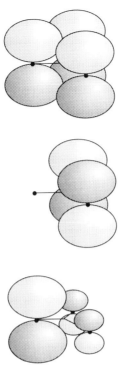

Fig. 4 The combinations of the 2p orbitals of the three oxygen atoms in a carbonate ion, CO_3^{2-}, from which molecular orbitals will be constructed. Only the top combination can overlap with a perpendicular 2p orbital on the carbon atom.

when their orbitals merge and a bond is formed when each species supplies one electron to the bond. This type of reaction is particularly important in a candle flame, which is sustained and propagated in this manner. Molecules are ripped apart in the hot tumult of the flame, and radical attacks radical, and radical attacks molecule. Incidentally, the propagation of combustion by radical chain reactions gives rise to a fourth way of extinguishing a flame: we can inject radicals into the flame; if they interfere with the normal propagation of the flame, then it will be extinguished. This interference is the basis of flame-retardants for fabrics, where the combustion of the fabric releases radicals that suffocate the flame.

Some reactions, though, proceed by a process in which a single species provides both of the electrons for the bond that is to be formed, and the orbital that carries them merges into an initially empty orbital on the second species. Such reactions are responsible, among others, for one possible future destiny of the CO_2 released by the flame: for CO_2 can act as an acceptor for an electron pair provided by an H_2O molecule, and hence CO_2 may become trapped into the lithosphere as calcium carbonate.

Whatever the process, we need to look for ways of achieving electron pairs. Faraday did not know that electrons exist. He would have been even more amazed to know that electrons have spin, and that electron spin is responsible for chemical bond formation. The stability of matter, the existence of a candle, the fact that when two lumps of matter touch they do not blend, is a deep consequence of the spin of an electron, and was not appreciated until the later 1920s. Turn off spin, and the world would collapse into a blob. There is no deeper remark I can make about the constitution of the world than that it is a collection of particles with half-integral spin bound together by the exchange of particles with integral spin.

But such analyses as I have described are crude relative to the detail in which some reactions are known. Thus, our understanding of some reactions relies on a knowledge of the response of electron clouds on one species to the approach of another species, and their detailed interaction as a bond is formed. In some reactions we can infer from experiment the details of the atomic arrangements that occur. Thus, we can make quite well-informed models of the atomic arrangements that take place when unburned hydrocarbon fragments from our flame escape complete combustion, rise through the stratosphere, and meet the vicious ozone (O_3) molecules that lurk there. We can go even further. We can investigate both computationally and experimentally the detailed waltz of atoms as they participate in reactions (Fig. 5). Here we are at the very heart of chemistry, the intimate moment of climax when an atom exchanges partners and reactants cross the Rubicon into products.

The return to life

Here it is appropriate to stand back from all this detailed molecular understanding, and capture Faraday's approach more closely by being more phenomenological. One of the modes of investigation that chemists employ to investigate chemical reactions is to establish the relation between the rate of a reaction and the concentrations of the species present. These so-called rate laws are differential equations. Now, differential equations are extraordinary. In 1865, at the end of Faraday's life, when this approach was first employed, the only rate laws that could be solved led to descriptions of relatively boring behaviour, such as the exponential decay of reactants and the smooth formation of products. Now, though, with computers we can investigate more complex rate laws. One of the solutions of a sufficiently complicated rate law is the elaborate pattern of a

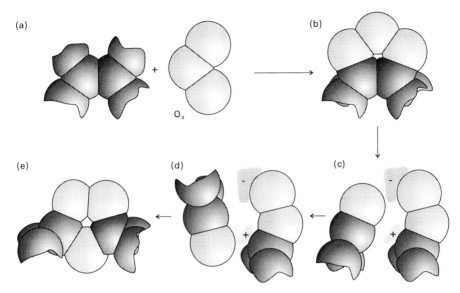

Fig. 5 The steps thought to be involved in the attack of an ozone molecule (O_3) on a carbon–carbon double bond.

peacock's plumage. Differential equations encountered in chemistry may be periodic in space and time, and lead to amazing patterns.

Although chemists cannot yet build peacocks, they can emulate their plumage. A model of the processes involved is the Belousov–Zhabotinskii reaction (Fig. 6), which was among the first to show spatial and temporal oscillations. Their work was scorned when it was first reported, but it has stimulated a central part of chemical research. It is easy to find many examples of spatially periodic chemical reactions in the natural world, all you need do is to inspect the coloration of animal pelts.

Even the turbulence of a candle flame is within the compass of these rate laws, for an oscillating system can break down into chaos. All this behaviour is embedded in the differential equations. Here chemistry is portraying the topological structure of the solutions of differential equations. Faraday would have been overjoyed that complex mathematics could be so vividly portrayed by matter.

Living candles

Now, though, it is time to return to life. As I remarked at the start, one of the happy aspects of a candle is that it bridges the inorganic and the organic, and thus is an anti-Medusa as well as a Medusa. Thus, the CO_2 released by our flame may end up either in rock, and be dormant for

Fig. 6 Some reactions show oscillations in time; some show periodic variations. This sequence of photographs shows the emergence of a spatial pattern.

aeons, or it may be incorporated into the much racier world of living things. Here, as Faraday well knew, it is the agency of green vegetation (and as we now know, chlorophyll) that captures CO_2 and its precious cargo of carbon and incorporates it into carbohydrates. Joseph Priestley, the inventor of soda pop and, incidentally, the discoverer of oxygen, knew that 'green plants could restore the air that had been injured by the burning of candles.' Faraday knew as much too, but he had no idea of the mechanism of the conversion.

In a sense, the progress in chemistry since Faraday's time is epitomized by the transition from a portrayal of photosynthesis in action, showing the bubbles of oxygen that emerge from green leaves, to Fig. 7, which is a

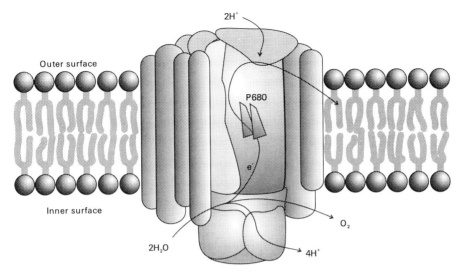

Fig. 7 Photosystem 2, the complex of molecules in the thylakoid membrane of chloroplasts that is responsible for trapping light and producing oxygen from water.

structural mechanistic illustration. Now we believe that we know in detail what happens within these extraordinary factories, which each year capture 100 trillion kilograms of carbon from the air. Chemists now know that a dramatic event takes place at the keystone of life: an electron is moved by a sunbeam. The energy stored in the photosynthetic unit is like a coiled spring that ejects an electron to a neighbouring molecule, whence it is whisked away and ratcheted out to the outer surface of the membrane. Meanwhile, the photosynthetic unit must prepare itself by gaining an electron so that it can act again. This it does by acquiring an electron from H_2O, so forming a proton and releasing oxygen.

Faraday lit his candle in the gloom of mid-nineteenth century London, and opened the eyes of his young audience to the current understanding of the composition and transformation of matter. He knew of atoms, but not of electrons. Mechanics was known, but quantum mechanics not. Now, when we light a candle, we can imagine the dance of its atoms and the ebb and flow of its electrons. Now our observations are enriched by insight of extraordinary intimacy. We can indeed see *deeply* into the workings of the world by the light a candle.

P. W. ATKINS, Ph.D

Born 1940, he was educated at the University of Leicester (B.Sc. 1961, Ph.D. 1964) in chemistry. He was a Harkness Fellow for the year 1964–

1965 at the University of California, Los Angeles where he carried out work on theoretical aspects of electron spin relaxation in liquids. He went to Oxford in 1965 as a Fellow of Lincoln College and University Lecturer in Physical Chemistry. He was awarded the Meldola Medal in 1969, and has been a visiting professor in Haifa, Sheffield, Tokyo, Grenoble, Shanghai, and Auckland. He was awarded an honorary doctorate of science by the University of Utrecht in 1992 for his contributions to chemistry. He has written a number of text books, including *Physical Chemistry*, *Inorganic Chemistry*, *General Chemistry*, *Molecular Quantum Mechanics*, and *Quanta*, and books on science for the general public, including *The Creation*, *The Second Law*, *Molecules*, and *Electrons, Atoms, and Change*.

Further reading

Faraday's lectures were originally published as Faraday, M. (1861). *A course of six lectures on the chemical history of the candle*. Royal Institution of Great Britain, London.

They are currently available, with additional experiments, as *Faraday's chemical history of a candle*, Chicago Review Press, Chicago (1988).

The material in this lecture, with directions to further reading, is set out at greater length in Atkins, P. W. (1991). *Atoms, electrons, and change*. Scientific American Library, W. H. Freeman & Co., New York.

Picture credits

Fig. 3. Photograph supplied by Sang-il Park and C. F. Quate, Stanford University.

Fig. 7. Based on an illustration by Govindgee and W. J. Coleman, in *Scientific American*, **262**(2), 50 (1990).

Painting with atoms

R. H. WILLIAMS

Modern developments in information technology and communications are familiar to us all and, in many ways, have transformed our way of life. These high technology developments are based around the silicon chip. Up to the 1950s, communication circuists were dependent on the thermionic valve, and the discovery of the transistor as a replacement for the valve was seen as enormous progress. However, in many ways the transistor could be considered as just a more convenient valve. It was really the development of the integrated circuit that led to really new technology and to the astonishing processing power of modern computers. During the past two decades the size of individual components on a microcircuit has been continually getting smaller, and the number of components per chip getting larger. The successful development of the 256 megabit memory has now been announced and it is likely that these circuits will be on the market in a few years' time. The cost of fabrication of modern integrated circuits, however, is extremely high and the cost per bit is now likely to increase. Further developments of such circuits may well, therefore, be now at the stage where they will be dominated by financial rather than technological considerations.

Present-day integrated circuits have individual components which are about 1 micron or 1 μm (or 10^{-6} m) in size, and circuits now being developed will have sub-micron components. Furthermore, many structures made for opto-electronics, including communication via optical fibres, are based on the fact that devices may be made ultra-small. This article will deal with the world of ultra-small semiconductor structures. Semiconductors are materials such as silicon (Si), germanium (Ge), gallium arsenide (GaAs) or aluminium arsenide (AlAs) which have electrical conductivity somewhere in between metals and insulators. The electrical conductivity can be controlled very precisely by 'doping' the materials with small amounts of an appropriate element (such as In or P for Si, and Si or Be for GaAs). It is of interest to consider how small one can make a semiconductor device, such as a diode, and what happens when a device is made forever smaller. The physics of sub-micron semiconductor structures is an extremely fertile field, which has led to many new kinds of devices. In this article I will attempt to transmit just a little of the flavour of this exciting field.

The behaviour of an oscillator, such as a pendulum, of about 1 metre in size is quite well known. The energy associated with such an oscillator is proportional to the square of the amplitude. For an oscillator which is ultra-small, less than 1 μm, the behaviour is quite different and is governed by quantum physics rather than classical physics. For a quantum oscillator the energy, E, is proportional to the frequency, ν, rather than the amplitude; in fact $E = h\nu$, where h is Planck's constant. This change from classical to quantum behaviour has far-reaching consequences.

Let us now consider the behaviour of semiconductor materials such as Si, GaAs, and AlAs in more detail. In all three semiconductors the atoms are held together by two electron bonds. These are normally tightly held in the coupling of the atoms and do not contribute to the electrical conductivity. However, if energy is supplied, for example in the form of heat or light, the bonds may be broken. The minimum energy needed to break a bond is called the energy gap, E_G, and is larger for AlAs than for GaAs or Si. If the bond re-forms, a corresponding amount of energy, E_G, may be liberated and for GaAs this may be in the form of light with a frequency of $\nu = E_G/h$, where E_G is the band gap of GaAs in this instance. This light, for GaAs, is in the infra-red part of the spectrum.[1]

Now consider the situation where a layer of GaAs is sandwiched between two layers of AlAs, or AlGaAs, as illustrated in Fig. 1. The energy gap for GaAs is less than that for AlAs or AlGaAs, resulting in a band profile as shown in Fig. 2. If the bond is broken the electron resides in the upper level (the conduction band), whereas if the bond re-forms the electron has energy corresponding to the lower level (the valence band). It is apparent that the upper energy levels lead to a 'potential well' in the GaAs. Electrons in the conduction band of the GaAs are confined in this two-dimensional potential well. The laws of 'quantum physics', remarkably, show that the minimum energy to break a bond in GaAs *increases* as the width of the potential well *decreases*. The electron (mass m) in the potential well has a minimum energy of

$$E_1 = \frac{h^2}{8\pi^2 m a^2}$$

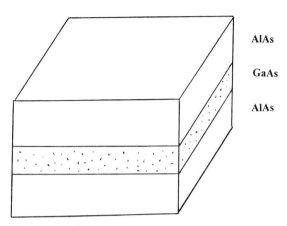

Fig. 1 Schematic diagram of a gallium arsenide (GaAs) layer sandwiched between aluminium arsenide (AlAs).

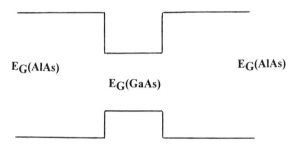

Fig. 2 Sketch of energy (vertical) as a function of distance (horizontal) for the AlAs–GaAs–AlAs sandwich. The energy, E_G is the minimum energy needed to break a bond and is smaller in GaAs than in AlAs. This results in a 'potential well' in the GaAs. In general AlGaAs is used as the high band gap material, rather than AlAs.

where a is the width of the well. Hence as a decreases the value of E_1 increases. Similarly, the light emitted when the bond re-forms is a function of the width of the potential well. Clearly, to ensure that the light emitted from GaAs is shifted towards the visible, the value of a needs to be reduced (typically to around 10 nm). This principle is fully illustrated in Fig. 3 and is used in the quantum well laser where the output wavelength may be controlled by appropriate choice of the quantum well width, a.[2]

Efforts are now under way to make laser structures with 'quantum wires' rather than quantum wells. Here the charge carriers are confined such that they can move only in one dimension in wires of typically 10 nm in thickness. Such lasers offer the advantage of reduced threshold

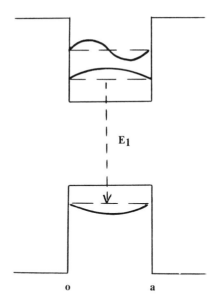

Fig. 3 Electrons confined in the GaAs potential well have allowed energies which depend on the width of the well. E_1 is the erngy emitted as light where a broken bond reheals. The energy E_1 increases as a decreases. Also shown are the form of the wave functions in a potential well.

currents but making wires of the required high quality is a considerable technological challenge. Much research is under way to fabricate 'quantum dot' structures where the electrons are confined in all three dimensions. These quantum wells, wires, and dots are referred to as 'low dimensional structures' and have nanometric dimensions. Nanostructures form the basis of nano-electronics which is set to become a technology of comparable importance to microelectronics.

By utilizing quantum confinement, laser structures are now in common use emitting in the visible part of the spectrum. For example, they are used in bar-code readers and compact disc players. Quantum confine has also been applied to make blue emitting lasers based on the semiconductors ZnSe and ZnS.[3] The achievement of 'blue' solid state laser structures is a long-held dream and offers interesting possibilities in colour display systems such as flat panel television. There will be very lively research programmes in this area during the next few years.

For quite fundamental reasons the semiconductor Si does not normally emit light when the broken bonds re-form. It turns out that the conservation of momentum in this case requires the participation of a 'phonon', or quantum of lattice vibration, and this is a process with a very low probability of happening. However, by chemically etching a Si crystal in such a way that it forms thin wire-like columns it has been discovered that Si will emit light very efficiently.[4] It is possible that this unexpected behaviour of what is referred to as 'porous silicon' is associated with the quantum confinement of electrons in the wire-like columns. There is much research under way to discover exactly how porous silicon emits light and how this may be utilized.

Structures of the kind shown in Fig. 1 provide the basis of the high electron mobility transistor (HEMT) which are the key elements in the amplifiers of satellite receivers. The HEMT responds well to very high frequencies and has a very low noise. The electrons in the HEMT move in the potential well in the GaAs and the current may be controlled via a gate electrode on the outside of the AlGaAs sandwich.[1]

How are ultra-thin layers such as the AlAs–GaAs–AlAs sandwich shown in Fig. 1 made? Several powerful methods are now available. One of these is 'molecular beam epitaxy' or MBE, illustrated schematically in Fig. 4. A substrate such as a GaAs crystal is enclosed in a vacuum chamber and heated so as to form a surface which is atomically clean. The chamber also contains crucibles known as Knudsen cells, or K-cells, which contain Al, Ga, and As. These cells are heated so as to generate molecular beams

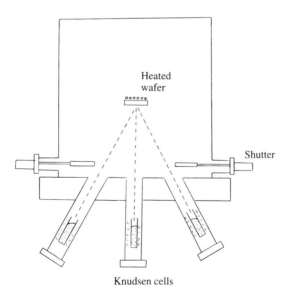

Heated
wafer

Shutter

Knudsen cells

Fig. 4 Sketch of molecular beam epitaxy apparatus for the growth of AlGaAs–GaAs–AlGaAs thin layers and quantum wells. Elementary As, Al, and Ga are evaporated from the heated Knudsen cells and fall on a heated wafer in ultra-high vacuum surroundings. A molecular beam may be switched off by pushing a shutter into the beam.

which fall on to the GaAs substrate. If all three beams fall on the GaAs a layer of AlGaAs is formed. If the Al beam is interrupted by a shutter only Ga and As fall on the substrate and a layer of GaAs is formed. By moving the shutter in and out of the beams, multilayers of AlGaAs–GaAs1–AlGaAs can readily be grown. The width of the GaAs layer is determined by the growth rate and the period when the shutter interrupts the beams. Modern MBE reactors have several K-cells, each with a shutter, and the growth may be monitored by the reflection of an electron beam from the surface (reflection high energy electron diffraction, or RHEED). The vacuum in the chamber has to be extremely high in order to avoid contamination of the layers during growth. Thin two-dimensional structures can be grown in this way. However, to generate quantum wires or dots or a complex circuit or structure it is normal to apply lithography and etching techniques which are well developed for structures of around 1 μm in size.[5] For much smaller structures it is customary to use the so-called 'electron beam lithography' approach where a finely focused high energy electron beam is used to generate the required pattern on a 'resist' for subsequent etching.[5] The etching step is often based on reactive ion beams.

The investigation of nanostructures often involve the use of high energy electron microscopes. These are large and expensive instruments, requiring considerable operator skill. There has existed a need for a much simpler technique to probe small structures and the invention of the scanning tunnelling microscope (STM) by Binning and Rohrer[6] in the early 1980s has revolutionized the situation. The principle of the STM is shown in Fig. 5. A metal tip is moved towards the sample surface by applying a voltage to the z-piezoelectric ceramic rod. The voltage between sample and tip leads to a tunnel current. When the tip is a few angstroms from the surface the z-movement is stopped and the tip distance, and therefore tunnel current, i_0, held at a constant value. The tip is now moved in the x and y directions by piezoelectric ceramic rods. If a step is encountered on the surface, as in Fig. 5, the distance d changes, leading to a change in tunnel current. The voltage on the z-piezoelectric ceramic rod is now changed so as to recover the original d and i_0. The voltages on the x, y, and z rods are displayed on a cathode ray tube and remarkably, not only are features such as steps observed, but individual atoms may be imaged. An example is shown in Figure 6 for the Si (111) surface.[7] This surface has been made atomically clean by heating in ultra-high vacuum and has a structure quite different from that in the bulk of the crystal. The surface is said to be reconstructed and in this case it displays what is known as a 7 × 7 form. The STM is now in use in many hundreds of laboratories world-wide and has been applied to generate striking images of a large number of solid surfaces.

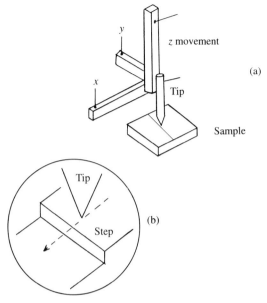

Fig. 5 (a) Schematic diagram of the scanning tunnelling micro-scope. The tip can be moved in the x, y, and z directions by appropriate voltages on the piezoelectric ceramic rods. (b) Illustration of the tip crossing a surface step (inset).

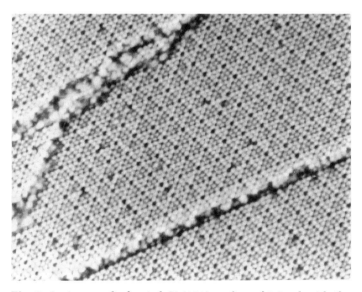

Fig. 6 An image of a heated Si (111) surface obtained with the scanning tunnelling microscope. Individual atoms are clearly seen.

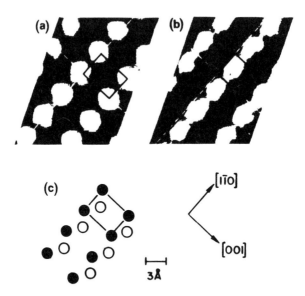

Fig. 7 A schematic diagram of the GaAs (110) surface is shown in (c) consisting of an equal number of Ga atoms (open circles) and As atoms (filled circles). The Ga and As atoms may be individually imaged by choice of tip-surface voltage. The Ga images are shown in (a) and the As atoms in (b). After Feenstra et al. (ref. 8).

Figure 7 shows the GaAs (110) surface imaged in two different modes.[8] In the first case the voltage on the tip of the STM is positive so that electrons tunnel from sample to tip. In this case the occupied surface states are imaged; these are electrons around the surface As atoms. If the polarity is reversed electrons tunnel from the tip into vacant surface states; these are s-like orbitals around the surface Ga atoms. The As and Ga atoms may thus be imaged independently simply by choosing the right voltage and polarity.

Other variants of the STM have been developed, and one of them, the atomic force microscope (AFM) has been particularly successful. In the AFM the image is obtained by measuring the physical displacement of a cantilever spring close to the sample surface. The displacement arises from Coulomb and van der Waals forces between the tip and the sample and the method can thus be applied to probe the surfaces of insulating materials. The AFM and STM can be used in various environments, including liquids, and the novelty of application seems almost endless.

I will now consider the application of the STM to probe ultra-small structures and also to make them. First consider the case shown in Fig. 8 where a single boron (B) atom is embedded in a Si (111) surface.[9] The tip

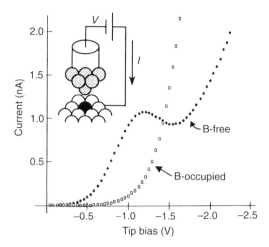

Fig. 8 Current–voltage spectrum for negative tip-bias taken in the region close to a single boron atom on a silicon (111) surface. The inset shows the geometry of the device. After Bedrossian *et al.* (ref. 9).

of the STM is then placed above the B atom and a voltage applied so that electrons tunnel between the tip and the surface. The current–voltage relationship for this B-occupied position is shown in Fig. 8. However, if the tip is shifted sideways slightly, the current–voltage spectrum is that shown for the B-free situation in Fig. 8. In one region it may be seen that the current decreases as the voltage increases, that is we have a structure or device showing negative resistance. Negative resistance is a central feature of oscillating circuits and here we have a negative resistance structure based around a single atom, indicating the potential for atomic-scale devices.

The STM can also be used to fabricate ultra-small structures. Consider the experimental set-up shown in Fig. 9(a) where the STM is used as an atom force microscope with a metallized tip.[10] A gold layer on a silicon crystal is coated with a PMMA film, between the gold layer and the tip. By application of an appropriate voltage it is possible to generate a tunnel current through the PMMA. The polymeric PMMA is an electron beam sensitive 'resist': it changes its properties under an electron beam. When the PMMA film is then etched in a suitable solution the regions exposed to the electron beam remain but those not exposed are etched away, leaving wire-like structures of PMMA, shown in Fig. 9(b). This is just one example where small-scale structures can be fabricated with the aid of the STM, and it is likely that the approach will be increasingly used in the future.

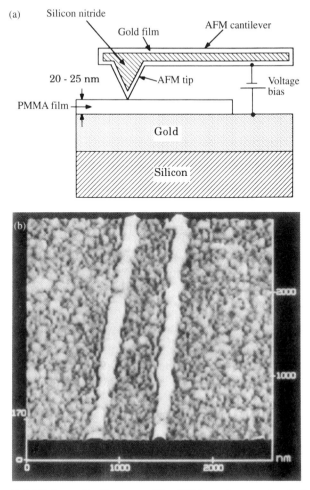

Fig. 9 (a) Schematic diagram of the STM used as an atomic force microscope for ultra-small-scale lithography using the electron beam between tip and gold substrate to modify the properties of the 'resist', denoted PMMA. (b) An image of two ultra-thin lines in the PMMA on gold. This was generated by etching away the regions of the PMMA not exposed to the electron beam. After Majumadar *et al.* (ref. 10).

Modern techniques, therefore, enable structures and devices to be fabricated with dimensions of the order of nanometres. The ultimate goal, of courses, would be to make structures by the manipulation of individual atoms. Astonishingly, the manipulation of single atoms and molecules to form interesting structures has been achieved using the tip of the STM. Two examples are shown in the last two figures, both from the laboratories of IBM in the USA. The first, Fig. 10 (upper left), shows the STM

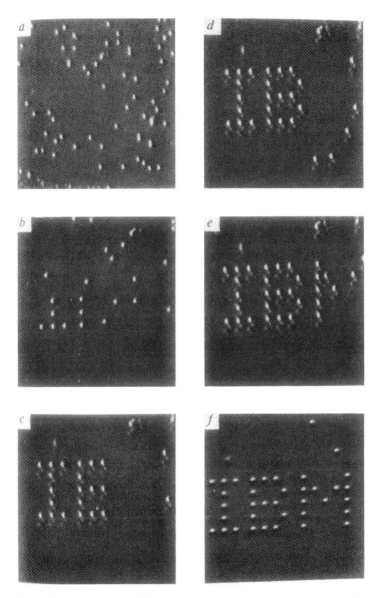

Fig. 10 A sequence of STM images for xenon atoms on the (110) surface of a nickel crystal. The xenon atoms have been moved with the tip of the STM to form the letters IBM. After Eigler and Schweizer (ref. 11).

images of xenon atoms on a nickel surface at low temperature.[11] Initially the atoms are randomly arranged. The tip of the STM was then placed above each atom and the small force between the xenon atoms and the tip allowed the atoms to be moved by moving the tip. In the last panel

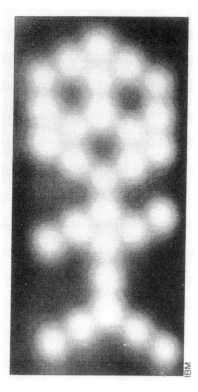

Fig. 11 A STM picture of carbon monoxide molecules standing
on end on a platinum surface, with the oxygen end upwards.
The molecules forming 'molecular man' were moved into these
positions with the tip of the STM. After Zeppenfeld (see ref. 12).

(bottom right) the atoms have been shifted to form the letters IBM. The
reader may wish to estimate the area that would be needed to store, let us
say, a large book in this way.

Finally Figure 11 shows a figure formed by moving carbon monoxide
molecules on a metal surface, using the tip of the STM. The molecules sit
with the molecular axis perpendicular to the surface. This arrangement
was described as 'molecular man', this one being the 'carbon monoxide'
version! Here the tip of the STM is being used like a paint-brush, an
example of nanometric scale 'painting with atoms'.

References

1. Morgan, D. V. and Williams, R. H. (1988). Advanced electronic materials for a
   new generation of microchips. *University of Wales Review*, **3**, 3–12.
2. Adams, A. and O'Reilly, E. (1992). Semiconductor lasers take the strain.
   Physics World, **5**, 43–7.

3. Gunshor, R., Nurmikka, A., and Kobayashi, M. (1992). II–VI semiconductors come of age. *Physics World*, **5**, 46–9.

4. Canham, L. (1992). Silicon optoelectronics at the end of the rainbow. *Physics World*, **5**, 41–4.

5. Morgan, C., Shan Chen, G., Boothroyd, C., Bailey, S., and Humphreys, C. (1992). Ultimate limits of lithography. *Physics World*, **5**, 28–32.

6. Binnig, G., Rohrer, H., Gerber, C., and Weibel, E. (1992). Tunnelling through a controllable vacuum gap. *Appl. Phys. Lett.*, **40**, 178–80.

7. Courtesy of W. A. Technology, Cambridge, UK.

8. Feenstra, R. M., Stroscio, J. A., Tersoff, J., and Fein, A. P. (1987). Atom selective imaging of the GaAs (110) surface. *Phys. Rev. Lett.*, **58**, 1192–5.

9. Bedrossain, P., Chen, D. M., Mortensen, K., and Golovchenko, J. A. (1989). Demonstration of a tunnel-diode effect on an atomic scale. *Nature*, **342**, 258–60.

10. Majumdar, A., Oden, P. I., Carrejo, J. P., Nagahara, L. A., Graham, J. J., and Alexander, J. (1992). Nanometer-scale lithography using the atomic force microscope. *Appl. Phys. Lett.*, **61**, 2293–5.

11. Eigler, D. M. and Schweizer, E. K. (1990). Positioning of single atoms with a scanning tunnelling microscope. *Nature*, **344**, 523–6.

12. Zeppenfeld, P. (1991). *New Scientist*, 23rd February 1991, p. 20.

ROBIN H. WILLIAMS, Ph.D., D.Sc., F.R.S.

Born 1941, he was educated at the University College of North Wales, Bangor, where he obtained an honours degree in Physics, followed by a Ph.D. After a period as a research fellow in Wales he was appointed Lecturer at the New University of Ulster, becoming Reader and then Professor. He was appointed Head of the Department of Physics and Astronomy in Cardiff in 1984, the post he still holds. His academic interests relate to the behaviour of semiconducting solids and in particular their surfaces and much of his research work is sponsored by industry. He has been Chairman of the Semiconductor Group of the Institute of Physics, and has chaired several committees of the Science and Engineering Research Council. He spent periods as visiting professor at Xerox Research Laboratories, IBM and the Max-Planck Institute in Stuttgart. He has been awarded the silver medal of the British Vacuum Society and the Max-Born medal and prize of the German Physical Society and the Institute of Physics. He is a Fellow of the Institute of Physics and of the Royal Society.

Superconductors: past, present, and future

PETER DAY MA, D.Phil., FRS

Not often does a discovery in basic science reach the front pages of the daily newspapers, but towards the end of March 1987, papers throughout the world carried a series of remarkable headlines: 'Electric Dream Has Come True', 'Scientists Herald Electronic Revolution', and 'The Woodstock of Physics' were just a few of them. Even more extraordinary was the fact that the event occasioning the headlines was a scientific conference, the kind of gathering of professionals not usually reported outside the staid pages of academic journals. What brought the world's TV and newspaper reporters to the Spring Meeting of the American Physical Society in Boston was an announcement that a series of novel chemical compounds had been discovered, which lost all electrical resistance at a much higher temperature than any previously known. The property of conducting electricity without resistance is called superconductivity, and the materials in question were dubbed high temperature (or, for short, high T_c) superconductors.

In this article I want to fill in the background that led up to what can only be called the hysteria that gripped the scientific world in 1987. I shall do so, first by exploring what superconductivity is and how it was first discovered, and then by describing how the high T_c materials came to be found, and the detective work needed to uncover their chemical composition and structure. Because I am at heart (as well as by training) a chemist, I shall also give a glimpse of the rich variety of other substances exhibiting this remarkable property, some of which (who knows?) may lead to advances as dramatic as those of 1987. The drama, of course, would not only be for science. The notion that electricity might be transported over long distances without expending energy is one to grip the imagination of engineers and technologists: that is why the news from the American Physical Society meeting made front page headlines. It is therefore pertinent to ask, six years later, how far those dreams have been fulfilled. First, though, we must go back nearly two hundred years, to consider how electricity is conducted by metals, and what the prefix '*super*' means in practice. Only then can the discoveries of the 1980s be placed in their proper context.

Perfect conductors and superconductors

One of the most important accomplishments of my distinguished predecessor as Director of the Royal Institution, Sir Humphry Davy, was to isolate several elements in groups 1 and 2 of the periodic table as a result of his work in electrochemistry. In fact, I like to tell visitors that about seven per cent of all the non-radioactive elements first faw the light of day at 21 Albemarle Street. Davy summarized many of his results in his Bakerian Lecture to the Royal Society in 1807 on 'The decomposition and composition of the fixed alkalies'.[1] After describing in that lecture some of the physical properties of what he called potash, but we would call potassium, Davy adds the bold statement: 'it is a perfect conductor of electricity'. Was he right? Sadly, no. Potassium is, of course, a metal and one of the better conducting metals at that, with a specific resistivity at room temperature of 60 micro-ohm cm. But that is certainly not zero. Indeed, much later in his career Davy returned to the general question of the conductivity of metals in a paper on '*The magnetic phaenomena produced by electricity*', published in the *Philosophical Transactions of the Royal Society* in 1821. Having determine, as he put it, that 'there was a limit to the quantity of electricity which wires were capable of transmitting', he conducted an exhaustive set of experiments on a variety of metals, designed to find out how the 'conducting power' depended on temperature, surface area, length, and (remarkably in view of its relevance to superconductivity) magnetic field. Summarizing the results of his experiments, Davy wrote: 'The most remarkable general result that I obtained by these researches, and which I shall mention first, as it influences all the others, was that the conducting power of metallic bodies varied with the temperature, and was lower in some inverse ratio as the temperature was higher.' What he had observed, therefore, was the general characteristic of metals that their conductivity decreases with increasing temperature.

Given that result, and with the benefit of hindsight, we can reapproach Davy's 1807 work on potassium and enquire if he might have been right if the temperature was lowered sufficiently. Again, sadly, the answer must be no. Figure 1 shows the resistance of a sample of potassium measured at very low temperature: the resistance indeed decreases as the temperature

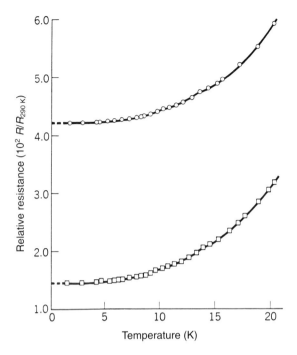

Fig. 1 The resistance of two samples of potassium at low temperature. (Reproduced from ref. 3).

is reduced but below 5° above absolute zero (5 K) it becomes more or less constant at about five per cent of its value at room temperature.[3] We know now that the electrons carrying the current, which behave as waves, are scattered when the atoms in the crystal lattice do not form an ideally equally spaced array. Such a situation can arise for two reasons. One is that real solids always contain impurities or occasional vacant lattice sites: the lower of the two curves in Fig. 1 was obtained from a purer sample than the upper one. Secondly, at any temperature higher than absolute zero the crystal lattice vibrates, so if we take an instantaneous photograph of the crystal the atoms will not be found on their ideal lattice sites. The latter is the physical origin of the phenomenon observed by Davy: as the temperature rises the lattice vibrates more, the electrons are scattered more, and the conductivity goes down.

After Davy's work the world had to wait another century before the first truly perfect conductor was found and, as is so often the case in science, it was an unexpected consequence of an apparently unrelated development. Throughout the 1890s and into the beginning of the present century a polite but intense professional rivalry existed between the laboratories of the Royal Institution and the Physics Department of the University of

Leiden in The Netherlands, personified by their two directors, respectively Sir James Dewar and Kamerlingh Onnes. Both were engaged in liquefying the rare gases, which required the production of very low temperatures. Dewar succeeded in every case except for helium, a final prize that fell to Onnes in 1908. It was therefore he, and not Dewar, who was in a position to start examining the properties of materials cooled to the lowest temperatures available at that time anywhere in the world. In view of the important part played by impurities and temperature in the conductivity of metals, Onnes decided to measure the resistance of a highly pure metal cooled to very low temperature and, because it could be purified by distillation, he chose mercury: the result (shown in Fig. 2, taken from his original publication on the subject) was extremely surprising.[4] The resistance decreased in the way characteristic of a metal as the temperature was lowered towards 4.2 K, but suddenly at about 4.3 K it dropped abruptly, becoming (as far as he could measure it) zero. The first perfect (or, as we now call it, super) conductor had been found.

Onnes' picture (Fig. 2) gives us a clue to several important aspects of superconductivity. First, the electrical resistance is not just low, it really is zero: current has remained circulating in suitably cooled superconduct-

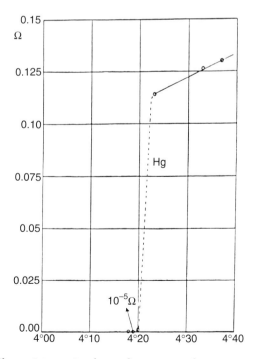

Fig. 2 The resistance in ohms of mercury at low temperature, as measured by Kamerlingh Onnes.[4]

ing wires for years without detectable degradation. Second, for every metal that exhibits it, the phenomenon occurs only below some characteristic temperature called T_c; for some metals (such as potassium, as it happens) it does not occur at all, even at the lowest temperatures accessible nowadays, numbered in millikelvins. A third matter, not apparent from Fig. 2, is that a superconductor is a perfect diamagnet. It was actually Faraday who first divided all substances into two classes, paramagnetic and diamagnetic, according to whether lines of magnetic flux (he called them 'magnetic conduction') arising from a uniform magnetic field bathing a lump of the material were drawn into it or expelled from it. The two cases are shown in Fig. 3, in a diagram taken from Faraday's paper, published in his *Experimental Researches in Electricity* in 1852.[5] In fact, a superconductor is a perfect diamagnet, with no lines of flux inside at all, in contrast to ordinary metals, which are weakly paramagnetic. However, that is only the case for magnetic field below some critical value, H_c, above which the material no longer superconducts, but reverts to a normal metal. Clearly, for any practical use that one may wish to make of superconductivity, H_c is equally important as T_c.

Below T_c and H_c our material is in quite a different state from that above these values; really a new state of matter. This is not the place to expound the theoretical explanation for the zero resistance and perfect diamagnetism, which was only established some 40 years after the original discovery, and is associated with the names of Bardeen, Cooper, and Schrieffer (called BCS theory for short).[6] Suffice it to say that in a normal metal the electrons move through the crystal lattice essentially independently of one another. By contrast in a superconductor they move as pairs (so-called 'Cooper pairs'). Agitating the lattice by heating it up, or applying a

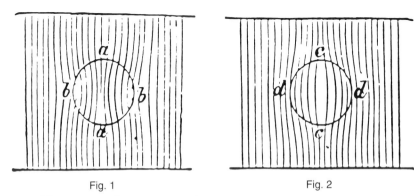

Fig. 1   Fig. 2

Fig. 3 Lines of magnetic conduction (*sic*) around a paramagnetic and a diamagnetic specimen, as drawn by M. Faraday, labelled respectively 'Fig. 1' and 'Fig. 2'.[5]

magnetic field, are two ways of pulling these pairs apart so that they revert to independent motion.

Exploiting superconductivity

The notion that wires can be made that would carry an electric current without any resistance is one to make an electrical or electronic engineer dream. But the dream can only be realized by keeping the wire at a temperature below T_c. Until six years ago, the only way to achieve the temperatures needed was to use liquid helium, still (80 years after its first liquefaction by Onnes) an exotic and expensive fluid. Not only is helium quite a rare element, produced in nature as a by-product of radioactive decay of heavy elements and hence found only in a few isolated deposits, but it has to be stored and transported in elaborately insulated containers. The latter, ironically, are called 'dewars'. Still, despite the complications and expense, superconductivity is such a unique property that it has found quite a number of applications, albeit of a specialized 'high tech' kind.

Many of the present day applications hinge on making magnets. In the Royal Institution it is scarcely necessary to dwell on the fact that when a current is passed through a wire formed into a coil, a magnetic field is produced at the centre; the bigger the current, the bigger the field, but in a normal metal the wire gets hot. Along a superconducting wire one can pass extremely large currents without it heating up provided only that the magnetic field generated by the coil remains below H_c. A coil made by Faraday, with about 200 turns on a cardboard tube, should produce a field of about 2000 G; one of comparable size made from a superconducting niobium alloy can reach about 30 times higher.

Very high field magnets first found application in physics research laboratories but, with the invention of nuclear magnetic resonance, which found wide acceptance as an analytical tool, they quickly came to be found in chemistry and biology laboratories. Companies such as Oxford Instruments sprang up to manufacture these complex and delicate items of equipment (Fig. 4). Even more widespread nowadays are the magnetic resonance body scanners found in many hospitals, based on super-conducting magnets large enough to insert a human frame into. What is believed to be the largest superconducting magnet ever constructed is shown in Plate 1. It was built at the Rutherfod Appleton Laboratory (RAL) three years ago for installation at the Large Electron–Positron collider (LEP) at the CERN High Energy Particle Physics Laboratory in Geneva. But even so large a solenoid has to be filled up inside with liquid helium.

Another important field of application for superconductors is to make

Fig. 4 A high field superconducting magnet and its cryostat. (Photograph by courtesy of Oxford Instrument Co.)

extremely sensitive detectors of minute variations in magnetic fields. When found in the earth's crust such changes might signal hidden oil or mineral deposits or, in the human body, what we might call 'brain waves'. The detectors in question are called SQUIDs (superconducting quantum interference devices), and consist of two layers of superconductor separated by an extremely thin layer of an insulator. Roughly speaking, the ability of electrons to travel through the insulator from one superconductor to the other is affected by small changes in the magnetic field of the surroundings.

It is against this background of important technology, hampered by the need to use elaborate cooling devices and expensive liquid helium, that the events of 1987, epitomized by the newspaper headlines I quoted, have to be seen.

1987: *annus mirabilis* for superconductors

After Kamerlingh Onnes' unlooked-for discovery of superconductivity in 1910, more and more metals were found to have this property at higher and higher temperatures. The league table of T_cs was headed successively by other elements, culminating in niobium at 9.3 K in 1930. After that binary alloys came into the picture and occupied metallurgists and materials scientists for many years, so that by the early 1970s the highest known T_c had crept up to 23 K. And there matters rested for 15 years. Learned theoretical papers were written predicting that T_c could never go higher than 25–30 K, and it is said (though I have not checked) that at that time candidates taking the theory option in the Cambridge Physics Tripos examination were asked from time to time to calculate the maximum T_c attainable under the BCS regime.

What a surprise, then, when at the end of 1986 a short article appeared in the German physics journal *Zeitschrift für Physik* under the unassuming title 'Possible high T_c superconductivity in the La-Ba-Cu-O system'.[7] The article was written by two Swiss scientists from the IBM Laboratory at Ruschlikon near Zurcih, George Bednorz and Alex Müller. They used the cautious word 'possible' because they had detected zero resistance but not, up till then, the other signature of superconductivity, diamagnetism. To a chemist, they also showed commendable caution in not identifying the formula of the compound concerned—in fact they did not know it. But the important measure was T_c; it was 30 K. After 15 years T_c had started to climb again.

Apart from the high T_c the other important characteristic of this new system was that it was not an alloy of the traditional kind known to metallurgists, but a ceramic. Oxide superconductors had been known prior to 1986 but with relatively low T_cs (e.g. $SrTiO_{3-x}$, 0.3 K; $BaPb_{0.7}Bi_{0.3}O_3$, 13 K; $LiTi_2O_4$, 13 K). Whatever the formula and structure of the new compound, the fact that it was an oxide placed it firmly within the realm of inorganic chemistry. Chemical analysis combined with X-ray diffraction, compared with measurements of the amount of diamagnetic material in specimens made by various methods in Europe, the USA, and Japan, showed very quickly that the optimum composition for superconductivity was $La_{1.85}Ba_{0.15}CuO_4$.

Consider this formula a bit more closely: if the compound had contained no Ba (always considered divalent by chemists), but only La (usually trivalent), and if oxygen had its normal valency of two, the copper would also have its customary divalency. Such a compound, La_2CuO_4, exists and is not a superconductor: it is not even metallic. On replacing a small fraction of the La^{3+} by Ba^{2+}, however, it becomes metallic and (for a fraction of Ba between about 6 and 10 per cent) superconducting. Under these circumstances the copper no longer has an integral oxidation state, but an average of say 2.15. Its valence shell contains a number of electrons that is neither integral nor a rational fraction. From work that we and others had done in the 1960s and 1970s, that kind of situation (called 'mixed valency' by solid state chemists) does lead to metallic conductivity, though it was not obvious why such a compound should superconduct. At once we decided to look into the finer details of the crystal structure and lattice vibrations, believing that these would hold the key to the mechanism.

At that time, the ceramic methods of preparation did not permit single crystals to be grown, so we worked with microcrystalline samples which, nevertheless, were beautifully crystalline over small volumes, as may be seen from the electron micrograph in Fig. 5. Most important to us was the precise positioning of the oxgen atoms around the copper, since that was the main factor determining the energy distribution of electrons in the conduction band. However, since they contain fewer electrons, oxygen atoms scatter X-rays much less than the other constituents of the crystal, hence diffraction experiments are best conducted using neutrons. On a wet Sunday afternoon in March 1987 we therefore found ourselves at the RAL at Chilton on the Berkshire Downs, in a small hut that houses the sample mounting and detectors of HRPD, the high resolution powder diffractometer. Some of the first data we obtained are shown in Fig. 6.[8] A reporter from *The Times* got wind of the fact that we were working on this novel material and interviewed me, leading to my only appearance on the front page of a national newspaper! The reason for the sudden press interest in our rather technical experiments was that news was emerging of an even more dramatic increase in T_c that was to change the landscape of superconductivity irrevocably.

It had just been found, rather surprisingly, that applying pressure to $La_{1.85}Ba_{0.15}CuO_4$ increased its T_c quite noticeably. A group at the University of Houston led by Paul Chu therefore decided to make a new compound by replacing La with the smaller Y. And lo and behold, T_c went up to more than 80 K![9] Ironically, like Bednorz and Müller, Chu did not know the formula of the compound his group had made, but a world-wide chase led quickly to the formula being established. The compound in question turned out not to be the analogue of the La,Ba one but had quite a different composition: $YBa_2Cu_3O_7$. Its structure too is quite different. The RAL group obtained a small powder sample and collected neutron diffraction data and we published a joint paper in *Nature* (Plate 2).[10] This was the material that led to the newspaper headlines with which I started.

The era of high T_c superconductors

When its composition is fully optimized, the material whose structure is shown in Plate 2, variously called YBCO or 123 for short, has a T_c of 93 K.

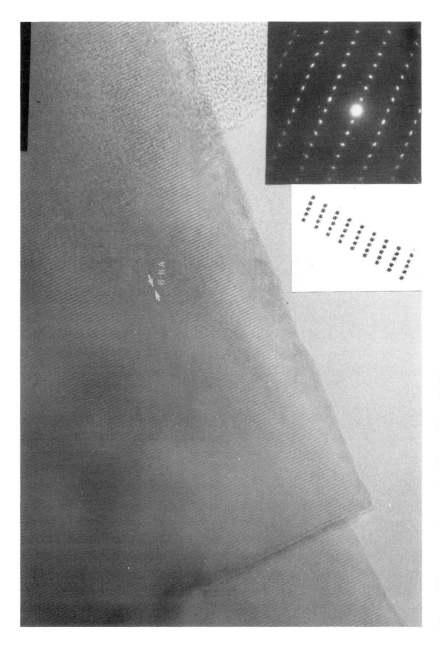

Fig. 5 Electron micrograph of $La_{1.85}Sr_{0.15}CuO_4$.

1. The largest superconducting magnet ever made.

2. The crystal structure of $YBa_2Cu_3O_7$[10].

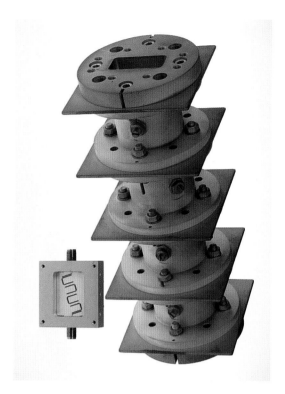

3. High-temperature superconductor planar thin-film microwave filter with a conventional waveguide filter (photograph courtesy of GEC–Marconi).

4. Crystal of an organic superconductor mounted for resistance measurements[15].

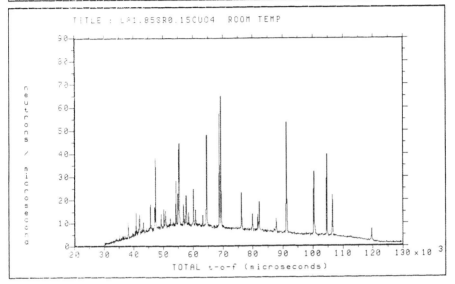

INSTRUMENT: HRPD
RUN NUMBER: 424
SPECTRUM :   1
LOCATION: HRPSDISK1:CHRPMGR.DATAJHRP00424.RAW

USER: MR/KP/PD
RUN START TIME: 21-MAR-1987 19:02:47
PLOT DATE: Sat 21-MAR-1987 20:59:36
BINNING IN GROUPS OF   10

TITLE : LA1.85SR0.15CUO4 ROOM TEMP

TOTAL t-o-f (microseconds)

Fig. 6 Raw neutron powder diffraction data on $La_{1.85}Sr_{0.15}CuO_4$ recorded on 21 March 1987.

The boiling point of liquid nitrogen is 77 K, so potentially at least exploitation of superconductivity in electrical and electronic engineering could be transformed. Not only is liquid nitrogen much easier to store and transport than liquid helium, it is much cheaper. In fact it used to be said that whereas liquid helium costs about as much as Scotch whisky, liquid nitrogen costs about the same as milk. But there are difficulties.

As already noted, the high T_c superconductors are not alloys but ceramics, a fact that has profound consequences for the way in which they are synthesized and fabricated. With regard to synthesis, conventional ways of making ceramics by heating mixtures of oxides together for long periods at high temperatures are certainly not conducive to forming material that can be made into wire. Also, the main characteristic of ceramics is brittleness and the high T_c compounds are no exception: a lump of YBCO has about the same mechanical strength as a tea cup! Grain boundaries, dislocations, and other defects also have an important effect on the key superconducting parameters such as T_c and H_c, and also the critical current density. Much ingenuity has been devoted to improving these properties, with results that may be seen from Fig. 7.

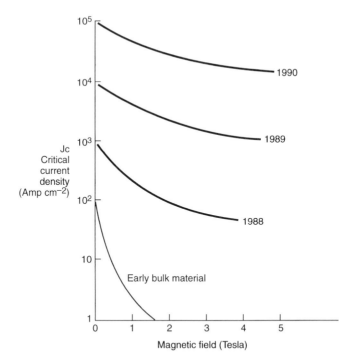

Fig. 7 Improvement in critical current density of YBa$_2$Cu$_3$O$_7$ samples from 1987 to 1990.

Whilst it is not possible to make wires out of ceramics alone, the friable and rather brittle material can be encapsulated in thin-walled silver tubes, which are then extruded at high temperature so as to trap the grains. Alternatively, it has been rolled into layers between sheets of silver and then cut up to form tapes, which can then be rolled into magnet coils. For electronic purposes, thin films have been deposited from the vapour phase by evaporating solid material, bombarding it with an intense laser beam. Another technique favoured by ceramicists for forming thick films is sol-gel processing. Organic salts containing the necessary Y, Ba, and Cu are hydrolysed to form a glassy homogeneous layer, which is then fired in a furnace to decompose it into the desired oxide.

Not only has a huge effort been made to optimize the properties and fabricate usable samples of YBCO but, in parallel, solid state chemists have continued their search for related materials that may have even more desirable properties. For example, about one year after the discovery of YBCO, Japanese workers prepared a series of phases in which the Y was replaced with Bi and the Ba with a mixture of Ca and Sr. Like the original La,Ba compound and YBCO, the Bi compounds (given the

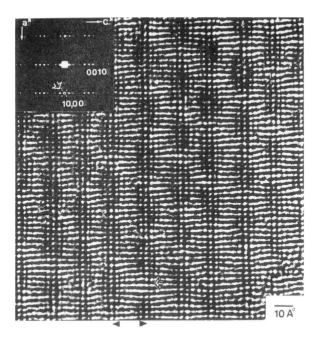

Fig. 8 Electron microscope lattice image of $Bi_2Sr_2CaCu_2O_x$. (Reproduced from ref. 11).

acronym BISCCO) contain layers of corner-sharing CuO_4 units, but present two extra features that are novel. First, there is not one but a series of compounds with different numbers of Cu,O layers sandwiched between layers of Bi and O, which gives each one a different T_c. Second, the Bi,O and Cu,O layers are not precisely flat, but have wavy corrugations. The electron microscope image in Fig. 8 shows a view parallel to the layers, clearly revealing the modulation which is brought about by a mismatch between the preferred Bi–O and Cu–O bond lengths.[11] Since, of necessity, they have to fit together in the lattice, a compromise is established in which first one and then the other becomes longer and shorter. This is an example of what physicists call 'competing interactions'. If the energy of the system is optimum for a variation in the lattice spacing that is not a rational fraction of the lattice repeat distance, the wavelength of the wavy corrugation (or modulation) is not an integral number of lattice spacings. Such structures, which are known in other inorganic compounds (though quite rare), are called 'incommensurate'. Physicists and chemists had a lot of fun working out the structures of the BISSCO family, although in the end the modulation seems to have little to do with the superconducting properties! However, because the BISSCO compounds have such a pronounced layer habit, they form better oriented films than YBCO, and hence have been tried out in several electronic applications (see below).

About a year after the BISSCO phases, T_c went up again with the discovery of closely related compounds containing Tl in place of Bi. The maximum T_c currently known in this series is 128 K but because Tl is extremely toxic it is unlikely that any serious efforts will be made to fabricate devices from it. Finally, just a month or two ago a German group reported a T_c of 133 K when the Bi or Tl is replaced with Hg. Again, given the toxicity of Hg, this discovery will most probably remain in the realm of basic science.

Applications of high T_c superconductors

In the six years that have followed the discovery of the copper oxide family of high temperature superconductors, many applications have been bruited. However, following the initial euphoria, quite a lot of obstacles have come up on the way to full technological exploitation, some of which were alluded to in the last section. Nevertheless, the first commercially available devices have started to appear in the last few months. Because it is so hard to make large homogeneous specimens of oxide superconductors, the first applications have been in small-scale electronics, especially in the microwave field. For example, GEC–Marconi recently put on the market a filter to separate microwave signals of different frequencies based on a one inch square film of the Gd analogue of YBCO deposited on a MgO substrate (Gd was chosen instead of Y because the repeat distance in its structure provided a better match to that of the substrate) (Plate 3). The device operates at 80 K, and therefore still requires cooling, but this can be accomplished by a small device that works on the Stirling cycle. Liquid nitrogen is not needed, only electric power. Even including the cooler, such a device is several times smaller and lighter than conventional ones made of copper.

Because miniaturization is possible, high T_c microwave devices are especially suitable for space applications. For example, trials of microwave antennae for installation in satellites are in progress in the USA, with 24 test components having been flown in spring of this year. A less demanding application of superconducting films is in shielding of sensitive electronic components from strong magnetic fields. Several Japanese companies have developed chambers lined with film for this purpose, up to 40 cm by 15 cm in size, inside which the magnetic background is reduced by a factor of no less than 10^5. Such boxes must surely constitute the ultimate in Faraday cages!

During the 1970s and 1980s, superconducting tunnel junctions related to the SQUID detectors described above were actively considered by several companies as potential circuit elements in computers because of the very fast switching times that can be achieved. The coming of high T_c materials revived interest in the topic, culminating most recently in an announcement by an American company, CONDUCTUS Inc., in February of this year. They have fabricated a 32-bit shift register (the most common element in information processing systems) from YBCO. Operating at liquid nitrogen temperatures, this digital electronic circuit runs at a clock speed of no less than 120 GHz, 1000 times faster than that of a current high speed personal computer. Incorporation of this device in commercial computers is confidently awaited.

Whilst large-scale uses of high T_c superconductors have not developed so quickly as those in microelectronics, there are one or two indications of viable applications. Current leads for superconducting magnets made of conventional materials are already on the market from Hoechst. The property being exploited here is not so much the superconductivity of the lead, but the low thermal conductivity of high T_c materials, because of their ceramic nature. In a rather more futuristic vein, it has been suggested that disks of high material might form the basis of frictionless magnetic bearings that could carry loads up to 100 kg at speeds up to 500 000 r.p.m. Such flywheels might form the basis of an energy storage system.

The future for superconductors

Only when the ceramic copper oxide superconductors burst upon the world did solid state chemists begin seriously to consider the synthesis of new superconducting compounds as a task falling within their field of competence, rather than that of physicists, metallurgists, and materials scientists. Nevertheless, some years before the *annus mirabilis* of 1987, two quite different series of oxides had been discovered, both of 'mixed valency' type and both belonging to structural types well known in solid state chemistry, that also behaved as superconductors. As noted earlier, they were respectively $LiTi_2O_4$ (1975), a spinel, and $Ba(Pb,Bi)O_3$ (1975), a perovskite, both with T_cs of 13 K. Since 1987, T_c in the Bi series has gone up to 30 K, with the preparation of $K_xBa_{1-x}BiO_3$.[12] There have also been tantalizing reports of zero resistance in other titanate phases at even higher temperatures, but they remain unconfirmed: chemical stability of defect oxide phases renders such work very difficult.

As long ago as the 1960s the American theorist, Bill Little, promulgated the idea of a new mechanism for superconductivity, which relied on coupling the electrons not to lattice vibrations but electronic excitations. Were substances to be found in which such a recipe could be realized, they would have T_cs measured, not in tens but hundreds of K. Little's proposal initiated a search especially among molecular-based compounds, the more usual hunting ground for chemists, and finally in 1980 the first molecular superconductor was discovered. The compound in question was a charge transfer salt, that is, one in which a planar organic molecule has lost an electron to form a cation, compensated in the crystal lattice by an equivalent number of negatively charged anions. The first organic superconductors were called Bechgaard salts after the Danish organic chemist who made them.[13] The organic cation was tetramethyltetraselenofulvalene (TMTSF for short), and the anions were small inorganic species such as ClO_4^- and PF_6^-. Their T_cs were low (1.3 K), but the important thing is that superconductivity had been carried into a completely new domain of chemistry. Subsequently, some 30 related compounds have been made, with T_c climbing to 13 K.[14] Of course, compounds of this kind are made in quite a different way from ceramics, in fact they are crystallized from an organic solvent electrochemically at room temperature. Nevertheless, wires are attached for resistance measurements in just the same way as to any other metallic conductors[15] (Plate 4).

The story of new superconducting compounds is far from over: just last year, one of my former graduate students, Matthew Rosseinsky, working as a postdoctoral fellow at AT&T Bell Laboratories in New Jersey, found that fullerene, the new form of carbon discovered only a year or two before, can also form salts that superconduct. This time they contain inorganic cations—in fact the element discovered by Davy, potassium! The first such salt was K_3C_{60} (T_c of 18 K) but the record in the series is now held by Rb_2CsC_{60}, with a T_c of above 30 K.[16]

Where will it all end? The next practical landmark on the temperature scale must be the temperatures that could be reached by thermoelectric cooling, say some 220 K. That is a long way from the present all-comers' record of 133 K, but remember those old Physics Tripos questions: the mechanism of high T_c superconductivity is still not agreed among theorists six years after its first discovery. So there is no a priori reason to rule out even higher T_cs, and when chemists become involved, the arena of substances to be investigated becomes dramatically enlarged.

On 26 May 1993, the Chancellor of the Duchy of Lancaster, Mr William Waldegrave, in his capacity as Minister for Public Service and Science, introduced the Government White Paper on Science and Technology, the first for 20 years. In it we find the following statement:[17]

> Basic research is undertaken to advance fundamental knowledge, irrespective of any foreseeable application. Such speculative research can be a major source of the revolutionary changes in our understanding of the world, producing important technological

advances. Many of the dramatic improvements in recent years in our quality of life and standard of living would not have been possible without these discoveries.

As we survey the history of electrical conductors, from Humphry Davy's 'perfect conductor' through Kamerlingh Onnes' low temperature research to the high T_c superconductors now beginning to find important applications, it would be hard to find a better example than superconductivity to illustrate the point that by finding unusual new properties of matter, new opportunities are created for improving technology and human welfare.

Acknowledgements

It is a pleasure to acknowledge Oxford Instrument Company, GEC–Marconi, ICI, and the Interdisciplinary Research Unit for Superconductivity in Cambridge for loaning material to illustrate this article, and Sir Martin Wood, Dr Nick Kerley, Professor Cyril Hilsum, Dr Neil Alford, and Dr Yao Liang for their help.

References

1. Davy, H. (1808). *Phil. Trans. Roy. Soc.*, **98**, 1.
2. Davy, H. (1821). *Phil. Trans. Roy. Soc.*, **111**, 431.
3. Macdonald, D. K. C. and Mendelssohn, K. (1950). *Proc. Roy. Soc.*, **A202**, 103.
4. Onnes, K. (1911). *Akad. van Wetenschappen* (Amsterdam), **14**, 113, 811.
5. Faraday, M. (1852). *Exp. Res. Electricity*, **3**, 204.
6. Bardeen, J., Cooper, L. N., and Schrieffer, J. R. (1957). *Phys. Rev.*, **106**, 162.
7. Bednorz, J. G. and Müller, K. A. (1986). *Z. Physik B*, **64**, 189.
8. Day, P., Rosseinsky, M. J., Prassides, K., David, W. I. F., Moze, O., and Soper, A. K. (1987). *J. Phys. C.*, **20**, L429.
9. Wu, M. K., Ashburn, J. R., Torug, C. J., Hor, P. H., Meng, R. L., Gao, L., Huang, Z. J., Wang, Y. Q., and Chu, C. W. (1987). *Phys. Rev. Lett.*, **58**, 908.
10. David, W. I. F., Harrison, W. T. A., Gunn, J. M. F., Moze, O., Soper, A. K., Day, P., Jorgensen, J. D., Hinks, D. G., Beno, M. A., Soderholm, L., Capone, D. W., Schuller, I. K., Segre, C. V., Zhang, K., and Grace, J. D. (1987). *Nature*, **327**, 310.
11. Gai, P. and Day, P. (1988). *Physica C*, **152**, 335.
12. For a brief overview of oxide superconductors, see Etourneau, J. (1992). In *Solid state chemistry: compounds* (ed. A. K. Cheetham and P. Day), p. 60. Clarendon Press, Oxford.
13. Jérôme, D., Mazaud, A., Ribault, M., and Bechgaard, K. (1980). *J. Physique Lett.*, **41**, L95.
14. See for example Saito, G. and Kagoshima, S. (1990). *The physics and chemistry of organic superconductors*. Springer Verlag, Berlin.
15. Doporto, M., Singleton, J., Pratt, F. L., Janssen, T. J. B. M., Perenboom, J. A. A. J., Kurmoo, M., and Day, P. (1992). *Physica B*, **177**, 333.

16. Fischer, J. A. and Bernier, P. (1993). *La Recherche*, **24**, 46.
17. *Realising our potential: a strategy for science, engineering and technology.* HMSO, London, 1993, para. 3.4.

PETER DAY, M.A., D.Phil., F.R.S.

Born 1938 in Kent, he was educated at the local village primary school and nearby grammar school at Maidstone. He was an undergraduate at Wadham College, Oxford, of which he is now an Honorary Fellow. His doctoral research, carried out in Oxford and Geneva, initiated the modern day study of inorganic mixed valency compounds. From 1965 to 1988 he was successively Departmental Demonstrator, University Lecturer and Ad Hominem Professor of Solid State Chemistry at Oxford, and a Fellow of St. John's College. Elected Fellow of the Royal Society in 1986; in 1988 he became Assistant Director and in 1989 Director of the Institut Laue-Langevin, the European high flux neutron scattering centre in Grenoble. Since October 1991, he has been Director and Resident Professor of Chemistry, The Royal Institution, and Director of the Davy Faraday Research Laboratory.

Sunlight, ice crystals, and sky archaeology

ROBERT GREENLER

Introduction

This article is about pretty things that can be seen in the sky, things that require no special equipment or location and so can be seen by anyone who looks. I have quite a long list of pretty things to choose from. The rainbow has received good publicity for a long time, and I could talk about the blue sky, the red sunset, black clouds, and the green flash. But I am not going to discuss any of those things here. I will restrict my comments to those effects that result from the interaction of sunlight with minute ice crystals tumbling through the sky. In fact, I will restrict my comments to the two kinds of ice crystals shown in Fig. 1. I will refer to the flat, hexagonal form as a plate crystal and to the long, columnar form as a pencil crystal (because of its resemblance to a wooden pencil—before sharpening). Both of these forms have been collected, photographed, and are known commonly to exist in the sky. They give rise to most of the observed ice crystal haloes and arcs.

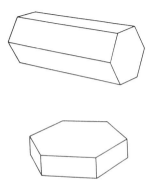

Fig. 1 The two forms of ice crystals that produce most of the observed ice-crystals haloes and arcs.

One of the things I hope to accomplish with this article is to acquaint readers with a whole class of beautiful effects that they may not have seen before, but which they will look for (and see) after this exposure. Lest the pleasure that comes from 'understanding' for its own sake is valued too lightly, recall that such an understanding of the world around us, and our excitement in that understanding, formed the very beginning of what we now call 'science'.

Haloes and sun dogs

The most common manifestation of ice crystals in the sky is the clouds composed of tiny ice crystals; for example, the high, wispy cirrus clouds seen in the sky on a warm sunny day. I will begin with the most common of the optical effects that results from ice crystals in the air: the halo that is sometimes seen around the sun or moon (Plate 5). Sometimes only a part of the halo is seen, while at other times an entire, bright halo is seen when the sky appears to be nearly clear and it might appear as if it contains few ice crystals. The halo is evidence that well formed ice crystals are present.

Figure 2 shows a light ray passing through a hexagonal crystal. The dotted lines extending the sides of the hexagon show that the ray passes through the crystal exactly as if it were passing through a 60° prism. It changes its direction (by refraction) both on entering and on exiting the prism. The resulting deviation of the ray through an angle of 22° is shown in the figure. This 22° deviation results from the 60° prism angle and the optical characteristics of ice. However, if the prism is rotated either clockwise or anticlockwise from the position shown, *the deviation increases!* The figure illustrates the minimum deviation angle, which occurs when the light ray makes the same angle with both faces of the prism; any other orientation of the prism will result in a greater deviation. One consequence

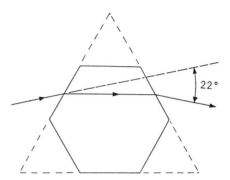

Fig. 2 Path of a light ray passing through alternate side faces of a hexagonal ice crystal.

of this minimum deviation angle is that if we have a large number of crystals, with different orientations, we would see that there are more rays deviated by approximately 22° than by any other amount. Small changes in the orientation of the prism, near this minimum-deviation orientation, do not change the deviation angle of the light by much.

Where would you look in the sky, filled with randomly oriented ice crystals that are illuminated by the parallel rays of light from the distant sun, to see an ice crystal that is deviating a light ray by an angle of 22° *and sending it to your eye* (only those rays that come to your eye give you any visual information about the surrounding world)? Figure 3 helps with the answer. You would look in a direction that is 22° away from the sun. That describes not one direction in the sky, but a circle around the sun. It is a circle with an *angular radius* of 22°; this is why the halo is called the '22° halo'. Different colours are deviated by different amounts in passing through a prism. Red light is deviated the least and, hence, the red circle is slightly smaller than the other colours and we usually see the halo with a red inner edge.

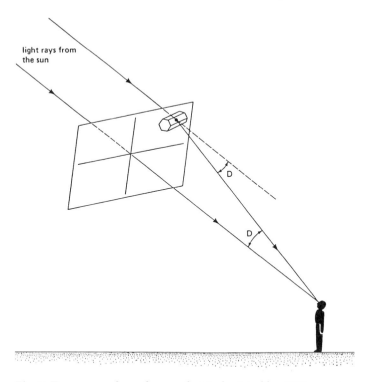

Fig. 3 To see a ray from the sun that is deviated by 22° in passing through an ice crystal, you look in a direction that is 22° away from the sun.

The 22° halo thus results from light passing through a 60° prism of ice. In the hexagonal crystals of Fig. 1, there are also crystal faces that make angles of 90° with each other; for example, in the plate crystal, each of the six side faces makes an angle of 90° with the larger, hexagonal faces. Light can also pass through a 90° prism of ice, which has a minimum deviation angle of 46°. The same kind of argument that explains the 22° halo would predict that there should also be a 46° halo. The 22° halo is big (44° across), but the 46° halo is really big, more than 90° from one side to the other. It is much less common than the 22° halo, but I think that even when it is present we may miss seeing it because we are not used to looking for things on so grand a scale. Plate 6 shows a very wide-angle photograph of the two haloes around the sun.

Plate 5 shows the (nearly) complete 22° halo. The light from the sides of the halo would be deviated to your eye by passage through the side faces of a plate crystal, for example, which is oriented so that those faces are nearly vertical. It will be helpful in describing the orientations of these crystals to define the crystal axis as a line passing through the centre of the crystal, at right angles to the hexagonal crystal faces. So, light from the sides of the halo in Plate 5 will get to your eyes by passing through the side faces of plate crystals oriented with their axes nearly vertical, whereas light from the top or bottom of the halo will be sent to your eyes by crystals with their axes roughly horizontal. In order to get the complete halo, crystals with all orientations must be present in the air. Sometimes, however, the ice crystals do not tumble randomly as they fall. A plate crystal made of ice, shaped like the model of Fig. 1, whose diameter is a fraction of a millimeter, will tend to become oriented as it falls in still air. The orientation is (perhaps surprisingly) such as to maximize the air resistance as it falls. A plate crystal of the appropriate dimensions falls flat, like a spread-eagled sky diver, with its axis nearly vertical.

How is the appearance of the halo affected by having a sky full of ice crystals, each with this orientation? The sides of the halo would be visible, but not the top or the bottom. And, in fact, this effect can be seen. Plate 7 shows a faint 22° halo, from a layer of randomly oriented crystals, and shows bright spots on either side of the sun, from a population of well-oriented plate crystals. Sometimes those spots appear surprisingly bright and are referred to in many folk sayings. They are commonly called 'false suns' or 'mock suns' or 'sun dogs'. Scientific folk sometimes call them parhelia.

The photograph in Plate 7 was taken north of the Arctic Circle, and that in Plate 8 was taken in Antarctica, at the South Pole. So, it is clear that sun dogs can be seen in the polar regions. I recently spent a year in Malaysia, living at 2° north latitude. I had not heard of any reports of sun dogs or haloes from Malaysia, but I have photographic evidence that they do occur there. High-altitude, ice-crystal clouds exist in the tropics and produce the same optical effects there as do ice-crystal clouds in the polar regions. But what about those of us who spend most of our time in more temperate climes?

Some years ago I had the pleasant experience of living in Norwich (England) for a year. I was working at the University of East Anglia in another area of my scientific interest, but I did bring with me a collection of pictures of rainbows, sunsets, and ice-crystal haloes. It happened that during the year I gave a number of talks about these pretty effects at different universities. In the question period following the first talk, one of the listeners framed his comment about ice-crystal effects in approximately this form: 'Well, Professor Greenler, these things are all very nice, of course, but you must realize that here in England, with all of our cloudy weather, we rarely have the chance to see any of them.' Although I do not remember how I responded to that comment, the next time it came up, I was prepared. (In fact, if the comment was not made, I brought the subject up myself.) My later response was something like this: 'Well, last evening, before I left Norwich, I looked at my diary where I had noted the days on which I saw sun dogs or some of the ice crystal haloes, and in the 23 weeks that I have been in England, I have seen sun dogs or haloes on 20 different days. So, I think that if you don't see them, it is because you don't look!'

Tangent arcs to the 22° halo

Frequently there is an intensification of the light pattern at the top of the halo, which you may have already noted in Plate 8. It usually takes the form of an arc that is tangent to (touching) the halo, although it sometimes curves away from the halo and sometimes towards it (see Plate 9). What is the cause of this arc? To direct light from the top of the halo to your eyes, crystals would have to be oriented with their axes approximately horizontal. Flat plate crystals do not fall with that orientation, but pencil crystals do. You can see the effect by throwing blades of grass into the air and watching the way they orient as they fall. This argument suggests that oriented pencil crystals might be responsible for the arc of light at the top of the halo, but how do we know whether that is really the case? We need to test that suggestion by finding some way to see what patterns of light would be caused by a large group of oriented pencil crystals. The computer can allow us to do this test.

In principle, it is not difficult to calculate how light changes its direction in going through an ice prism. There is a mathematical expression

(Snell's Law) that gives a simple relationship between the angle a ray makes with the surface on the air side of the surface and that on the ice side of the surface. We have only to apply Snell's Law to the light ray as it enters the ice crystal and again as it leaves. We use three angles to specify the orientation of the ice prism, then we calculate the direction of a light ray from the sun after it passes through the prism, and finally calculate the direction in the sky from which a crystal with that orientation would send the light ray to your eye. On a piece of paper the computer then plots a point, representing the location of an ice crystal that would 'light up' as its light comes to your eye. If this calculation is repeated a hundred thousand times for a hundred thousand different ice-crystal orientations, the resulting dot diagram shows us the light pattern in the sky that would be produced by the collection of ice crystals.

We started off with a simple distribution of orientations: we let the crystal axis be exactly horizontal, with the crystal rotated by a randomly chosen angle about its crystal axis, and its axis pointing in a randomly determined direction in the horizontal plane. We considered only rays that passed through alternate side faces as shown in Fig. 2.

Typical results are shown on the left side of Plate 9 for four different elevations of the sun above the horizon. The horizon is represented by the heavy horizontal line. The inner edge of the 22° halo is represented by the circle. The dots show the light pattern that we predict for this collection of ice crystals. For the sun on the horizon (elevation of 0°), the predicted V-shaped arc matches quite well the arc on the adjacent photograph with the sun near the horizon. As the sun elevation increases, the upper tangent arc flattens. At a sun elevation of 20°, the prediction matches well the photograph with the sun at the same altitude. At 30°, the arc is predicted to be a nearly horizontal band (where it has great enough intensity to be seen). The adjacent photograph shows only one arc; the 22° halo is not visible. Apparently all of the ice crystals were oriented pencil crystals, giving no trace of the circular halo, but producing only the horizontal band, in fair agreement with the computer simulation. Notice that the simulations predict a lower tangent arc at the bottom of the halo. In the first two examples, this arc would be below the horizon, but could be seen from an aeroplane flying over an ice cloud.

When the sun is at an elevation greater than about 40°, an interesting change takes place; the upper and lower arcs join at the sides to form a complete halo, completely encircling the 22° halo. This is called the circumscribed halo and is shown in the simulation of 50° sun elevation in Plate 9. Fraser's adjacent photograph shows the corresponding reality, produced by the laws of nature.

We have many more photographs of the upper and lower tangent arcs and of the circumscribed halo for a wide variety of sun elevations, and we have computer simulations that match them quite well.[1] What do we conclude from all of this? The exercise is really a rather nice illustration of how the scientific method (sometimes) works. We saw these arcs in the sky and wondered what caused them. We made the hypothesis that they might result from light passing through oriented pencil crystals. Based on that idea, we used our hypothesis and the computer simulation to predict how the shape of the arcs would vary with the sun elevation and how they turn into the circumscribed halo at high elevations. The impressive agreement of these predictions with the data (i.e. the photographs) convince us that we have indeed found a good explanation for the beautiful arcs.

Arcs near the 46° halo

I will show you some other arcs that result from light passing through the 90° prism faces of these hexagonal crystals. The presence of sun dogs gives us evidence that sometimes there is a large population of plate crystals oriented with their axes nearly vertical. What effects would result from light taking the path as shown in Fig. 4 through crystals with such orientations? The computer-simulation approach is a way to answer just that sort of question. The answer is shown on the right side of Fig. 4. In

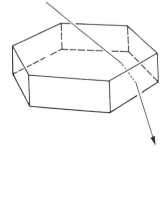
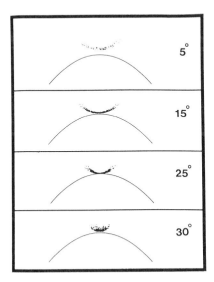

Fig. 4 On the left is shown a ray path through the crystal. To the right is the predicted effect (circumzenithal arc), for different sun elevations, resulting from such rays passing through a collection of plate crystals with nearly vertical axes and with random orientations around those axes.

that diagram, the curved line represents the inside edge of the 46° halo
and the dot diagram is the arc in question. It is difficult to look at a flat
picture and to know how to map it on to the dome of the sky. I can
describe precisely in words the position of this arc: it lies on a circle, the
centre of which is the point that is directly overhead. That point is called
the zenith and the name of this arc is a good description of its form: it is
called the circumzenithal arc—literally, an arc that runs around the
zenith. As the sun is higher in the sky, the circle of this arc gets smaller
and it only occurs when the sun elevation is less than 32°.

From our success in simulating the shapes of the upper and lower
tangent arcs to the 22° halo, and their evolution into the circumscribed
halo, we have confidence that the oriented pencil crystals really do exist
in the atmosphere. Suppose we trace rays taking paths labelled A and B
(Fig. 5) through a collection of pencil crystals with horizontal axes, but all
rotations about those horizontal axes, and with those axes pointing in all
possible directions in the horizontal plane. Our simulations (Fig. 5) show
that rays taking the 'B' path produce the upper arc (supralateral arc) while
the 'A' rays produce the lower arcs on either side of the 46° halo (infralateral arcs). These infralateral arcs, along with a number of other effects
that I have discussed, are shown in the photograph of Plate 10.

There are many more effects that I could show you that arise from light
interacting with these two simple ice-crystal forms.[1] However, I want to

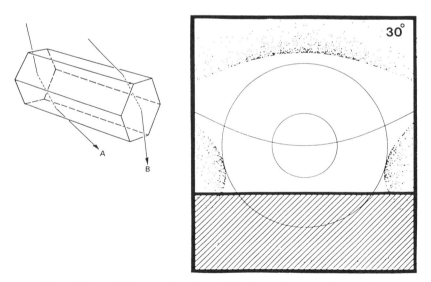

Fig. 5 Rays taking paths A and B through pencil crystals,
oriented with horizontal axes, produce the supralateral arc near
the top of the 46° halo (ray B) and the infralateral arcs at the sides
of the halo (ray A) for a sun elevation of 30°.

5. The 22° halo, photographed by the author in Wisconsin.

6. The 22° and 46° haloes, photographed by Alistair B. Fraser.

7. Self-portrait of the author, pointing toward sun dogs, near Point Barrow, Alaska.

8. Sun dogs and other effects, photographed by the author at the US Antarctic research station located at the South Pole.

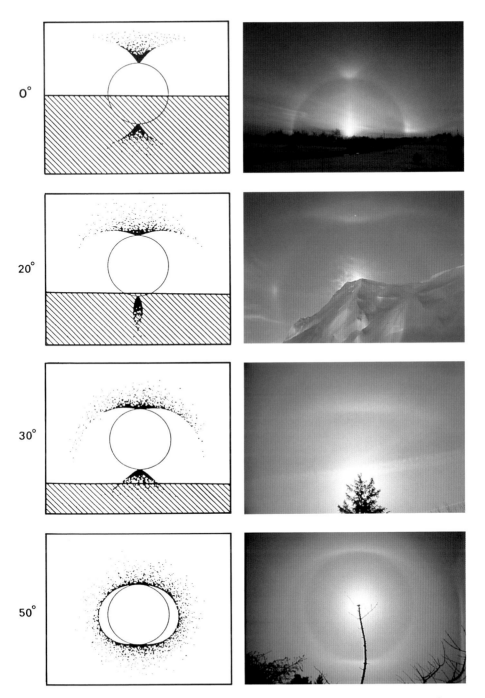

9. On the left are computer simulations for the specified elevation of the sun above the horizon. The photographs on the right were taken by (in order from the top) James Mallmann in Wisconsin, Evan Noveroski in Antarctica, the author in Wisconsin, and Alistair B. Fraser in the state of Washington.

10. Complex ice-crystal display, showing an infralateral arc, photographed by Takeshi Ohtake in Alaska.

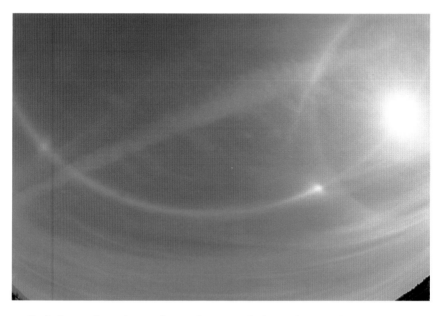

11. Parhelic circle with sun dog and 120° parhelion, along with the 22° halo and its upper tangential arc, photographed in New Mexico. (Courtesy of the Los Alamos National Laboratories.)

get on to discuss a famous report of a complex optical display that appeared in the skies of St Petersburg, Russia, in 1790.

The St Petersburg display of 1790

On 18 June 1790, Tobias Lowitz saw an amazing collection of haloes, streaks, arcs, and spots in the St Petersburg sky. His drawing of what he saw was published[2] in 1794 along with a brief commentary on the appearance of the effects, but with no explanation of their origins (see Fig. 6). Since then, people who have looked into the role of air-borne ice crystals in optical manifestations of the sky, have recognized some of the elements of that display. In fact, it has drawn the attention of those of us who try to understand such effects—as the test of our understanding. While some of the elements are easy to understand, others had eluded all efforts of explanation. Once we had achieved some success with our computer-simulation method in explaining optical effects that had pre-viously escaped understanding, we were drawn, as a moth to the candle, to the St Petersburg display.

In the centre of the drawing, a point is labelled zenith and on the four sides are compass directions. It seems clear that Lowitz is representing a display that covers the entire sky and is representing it in a wide-angle view that has the zenith in the middle and the horizon all around the edges. We decided to try to use our new-found art to simulate the entire display. How do we map the hemisphere of the sky on a two-dimensional plane? (That is a question that map makers have wrestled with ever since they concluded that the earth is spherical.) We decided, rather arbitrarily, to use the same mapping as is done by a fisheye lens (a very wide-angle lens that records an entire 180° field of view within a circle on the film). We chose a sun elevation of about 50° as a match for the main features of the drawing.

Several of the features of Lowitz's report can be understood from my earlier discussion. Randomly oriented crystals produce the 22° and 46° haloes. The circumscribed halo for a sun elevation of 50° (see photograph and simulation in Plate 9) is clearly a prominent part of the display. Infralateral arcs are strongly represented. The sun dogs in the drawing have moved away from the 22° halo—and we understand this effect as the sun is high above the horizon. The crystals have vertical axes and, for high sun elevations, the light rays come through these crystals at a steep angle. For such rays the effective angle of minimum deviation is greater than 22° and the sun dogs move away from the halo. The effect is shown in the wide-angle photograph of Plate 11. That photograph also shows the large circle that passes through the sun and the sun dogs and circles around the

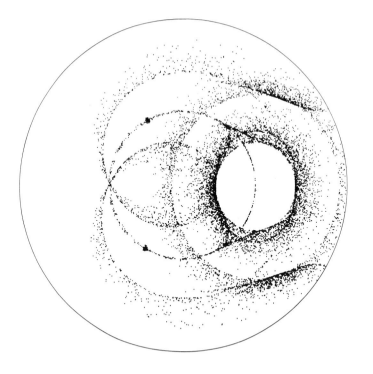

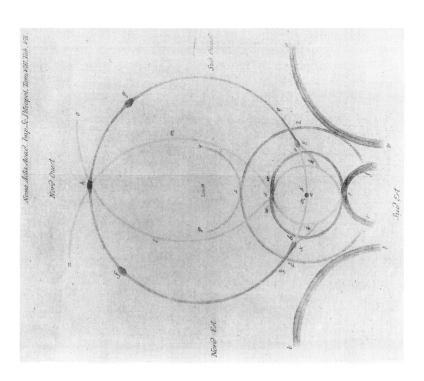

Fig. 6 On the left is the drawing by Tobias Lowitz from reference 2; on the right, a fisheye projection of the simulation of Lowitz's display for a sun elevation of 50°.

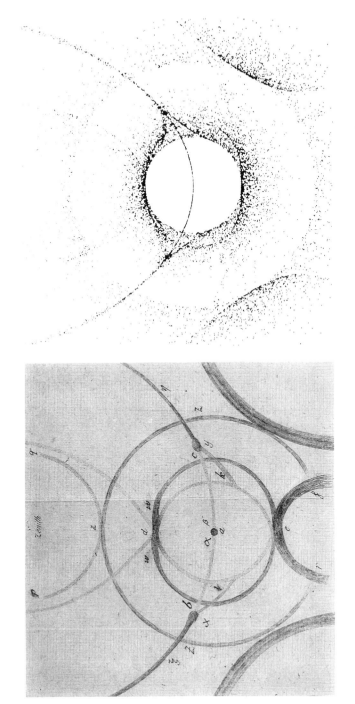

Fig. 7 A portion of Lowitz's drawing, matched by the simulation plotted with the perspective of an ordinary photograph (or eighteenth-century painting).

sky at a constant elevation above the horizon. It is called the parhelic circle and results from light reflected off an array of vertical reflecting planes filling the sky. There are two sources of such reflecting planes: the side faces of oriented plates and the end faces of oriented pencils.

The photograph of Plate 11 also shows one of the two bright spots on the parhelic circle (marked 'f' and 'g' by Lowitz). We can reproduce these spots with light rays that enter the oriented plate crystals and experience two internal reflections from the crystal faces before emerging. They are called 120° parhelia, because they are 120° in azimuth (angles measured in the horizontal plane) from the sun. Another ray path involving one internal reflection reproduces the arc touching the top of the 22° halo with its arms crossing at the parhelic circle, opposite the sun.

The two little arcs stretching from the sun dogs down to the 22° halo are rare, seldom-photographed arcs that take their name from their occurrence in this drawing. They are called Lowitz arcs and their origins have long been debated. We can reproduce them with rays that pass through alternate side faces of plate crystals with a peculiar set of distributions. Think of an axis passing through the opposite points of the hexagonal face of a plate crystal. Keep that axis horizontal and let the crystal spin about it. That may seem to be an improbable event until I show you that one of the ways a flat plate can fall—in air—is to rotate about a long axis! Do it yourself by holding an ordinary playing card horizontal by thumb and finger at a long edge. Let it roll off your finger as you release it, and you will see it spin about its long axis as it falls.

All of these things are included in the simulation shown next to Lowitz's drawing in Fig. 6. We find that we get a better agreement with the shape of the effects near the sun, for example the infralateral arcs, by a different kind of mapping. Figure 7 shows the simulation plotted with the perspective of an ordinary photograph or that of an 18th century painter painting a normal scene. Compare it with the corresponding part of Lowitz's drawing in the same figure. The improved agreement suggests to me the first of a series of insights as to what Lowitz did on the day he observed this display. It seems that he first drew the effects in the vicinity of the sun, with the natural perspective one would use to represent a small area of the sky, and then, somehow, added all of the other things—some of which were located behind him—to get them all recorded on the flat piece of paper.

Now look at the drawing and simulation in Fig. 7 to see how well we are doing. Several features of the simulation are quite reasonable matches to Lowitz's drawing, but there are some problems. In the drawing, the circular arc at the top of the 46° halo is most certainly the circumzenithal arc. Why do we not show it on the simulation? Because that arc does not occur when the sun is higher than 32° above the horizon! What about the arc that turns downwards from the 22° halo? We tried everything we could think of to get such an arc for a sun elevation of 50°—and failed. We could get a lower tangent arc (or a near relative of that arc called a Parry arc) that would give such an effect for a sun elevation of 30°, but not for the 50° elevation of this picture. Then we noted another clue to the puzzle. There are 90° between horizon and zenith. The 46° halo has an (inner) diameter of 92°, and yet Lowitz shows it fitting between horizon and zenith with space to spare, both above and below! Then we began to understand. Lowitz says that he first saw the display at 7:30 in the morning (when the sun would have been low on the horizon). He describes the display as having attained its most beautiful perfection at 10 o'clock in the morning (when the circumscribed halo would exist as shown in his drawing). But at other times he saw other arcs, and it finally became clear to us that, although he does not explicitly say so, he added these things, *from differ-ent sun elevations*, to this master drawing. He shows the circumzenithal arc as it would have appeared when the sun was about 20° above the horizon, and the lower arc, touching the 22° halo, when the sun was about 30° high. Once we understand this, we realized that his notes confirmed it. At 7:30 in the morning he describes the arcs (of the circumscribed halo) as not yet in their perfection but slowly developing from the brilliant light at the top of the halo until at 9 o'clock the entire halo was formed. He describes the evolution of other arcs, which our simulation reproduces, as the sun rises.

So, more than two centuries after Lowitz recorded this amazing display, we can understand what he saw and described, but did not understand. And in our process of understanding, we get some insight into the activities of the man who stood, looking at the skies of St Petersburg on that 18th day of June, and marvelled at the beauty he saw displayed there. Don't you think that's interesting?

References

1. See Greenler, R. (1980). *Rainbows, halos, and glories*, Cambridge University Press, and references therein. (Paperback edition published 1989.)
2. Lowitz, T. (1794). *Nova Acta Academiae Scientiarum Imperialis Petropolitanae*, **8**, 384.

ROBERT GREENLER

Born 1929, he was educated at the University of Rochester (B.S.) and Johns Hopkins University (Ph.D.). Employed in the Research Division, Allis–Chalmers Manufacturing Company, 1957–62 before joining the faculty at the University of Wisconsin–Milwaukee. Has been instrumental in the development of the Laboratory for Surface Studies at Milwaukee, which involves people from several academic disciplines. Major areas of research concern the study of solid surfaces and molecular interactions that take place at surfaces, optical effects of the sky, and the iridescent colours seen in many biological organisms. Was Senior Visiting Fellow, 1971–2, in the School of Chemical Sciences, University of East Anglia, Travelling Lecturer for the Optical Society of America, 1973–4, and has been elected a Fellow and Director-at-Large of that society. Was Senior Fulbright Scholar at the Fritz Haber Institute of the Max Planck Society in West Berlin, 1983. Was at the US Research Station at the South Pole in Antarctica in 1977 studying optical sky phenomena and the ice crystals that cause them and continued that study in the Arctic at Point Barrow, Alaska in March 1978. Received a UWM Foundation award for excellence in research in 1980. Elected a Fellow of the AAAS in 1983, President, Optical Society of America, 1987, member of the governing board of the American Institute of Physics, 1986–8 and a member of the Council of Scientific Society Presidents, 1986–8. He has an active interest in science education and is organizer and director of *The Science Bag* a series of public science programmes at the University of Wisconsin–Milwaukee. In 1988 received the Millikan Lecture Award from the American Association of Physics Teachers 'for creative teaching of physics' and spent 1990–1 in Malaysia teaching in a cooperative programme organized by a consortium of midwest universities and the Malyasian Government. In 1993 received the first Ester Hoffman Beller award given by the Optical Society of America for '... extraordinary leadership in advancing the public appreciation and understanding of science...'.

Supernovae: the deaths of stars

PAUL MURDIN

The supernova of 1054

Imagine that you are an astronomer of 950 years ago, living in what is now called Arizona. You are not only an astronomer, you are also a historian of your people, the Anasazi Indians; you are a genealogist, an archivist, and the keeper of the calendar. Each morning you go to a special place in Chaco Canyon to observe the Sun rise. It is the azimuth of the Sun on the eastern horizon, moving seasonally to the north and south against the natural features, which tells you the date of the year. This enables you to declare the time to plant seeds so that the crops may be harvested in due course. Your ancestors have taught you the regularities of the sky and the positions of the brighter stars. The cycles of the planets like Venus and Mercury have become completely familiar.

Contemplate therefore your surprise on 4 July, AD 1054, as you mount the rocky platform—your solar observatory—from which you regularly make your observations. In the dawn sky you see not only the crescent moon, but also, off its southern tip, a bright star, unexpected, never before seen—a jarring anomaly in an otherwise perfectly familiar cosmos. You have observed a nova, a new star which has appeared where no star was noticeable before, in fact a supernova, an especially bright example, representing the explosive death of a star. You are so moved by this observation that you record the scene in a pictograph, still visible high on the rocky face of the now-collapsed observing platform. The pictograph comprises a crescent moon, a bright star, and, by way of a signature, your handprint (Plate 12).

The scene which I have imagined in order to explain the Chaco Canyon pictograph of the prehistoric Pueblo Indian people is developed from fragmentary evidence from both archaeology and astronomy. It would be a brave man who would swear to all its details, interpreted from the circumstances. The astronomical details of the view from Chaco Canyon on 4 July 1054 are, however, relatively secure. The appearance in the sky of a supernova, south of the crescent moon, is vouchsafed by written records.

One record is an Islamic medical textbook of the thirteenth century. *Important information concerning the generations of physicians*, by Ibn Abi Usaybia, written in 1242, refers to epidemics in the Middle East, in cities such as Damascus, Cairo, and Constantinople, as brought about by the appearance of a new star in the zodiacal sign of Gemini in 1054 (Fig. 1). We maintain a linguistic fossil of the idea that diseases can be caused by celestial events, in the word 'influenza'. The Italian word 'influence' is used for the particular disease which swept Italy in 1743, provoked as it appeared at the time by the arrival of a comet.

The most detailed accounts of the supernova of 1054 appear in oriental records from China, Japan, and Korea. To determine the direction from which the 'cosmic wind' would blow their kingdoms, oriental emperors maintained a civil service of astrologers to record and interpret astrological signs. Thus extensive series of observations were made of astronomical and what we would now call meteorological phenomena. This costly activity was maintained so that an emperor could be warned, for example, of impending invasion or insurrection, one interpretation of the appearance of what the Chinese astrologers termed a new 'guest star'. Being a civil service, the astrologers kept detailed records. Some records survived the destruction following the invasions of the Mongol barbarians and have come down to us, a rich mine of astronomical data.

From these records we know the time of appearance of the supernova of 1054. ('In the first year of the period Chih-ho, the fifth moon, the day chi-ch'ou, a guest star appeared southeast of Thien-kuan' is the statement in *Sung Shih*, the official history of the Sung dynasty by the fourteenth century historians Toktaga and Ouyang Hsuan. This date is July 4 1054.) We know that the supernova of 1054 lasted for 630 days, since Chang Te-hsiang writes in the *Sung hui-yau* that the Director of the Astronomical Bureau reported in 'the third month in the first year of the Chia-yu period' (on about 1 April 1056) that it had become invisible.

Astronomers can reconstruct the outline of the light curve of the supernova of 1054, since, according to the Director of the Astronomical Bureau it was originally so bright as to be visible in daylight, like Venus, before it faded into the evening sky two years later.

The Crab Nebula

We also know the direction in which the supernova of 1054 appeared. The Chinese astrologers divided the stars into smaller constellations than

الخلفاء المصريين وجرت بين ابن بطلان وابن رضوان وقائع كثيرة في ذلك الوقت ونوادر
نظر رفعة لا تخلو من فائدة وقد تضمن كثير من هذه الاشياء كتاب الغاز ابن بطلان بعد
خروجه من ديار مصر واجتماعه بابن رضوان ولابن رضوان كتاب في الرد عليه وكان ابن
بطلان أعذب ألفاظا وأكثر طرفا وأميز في الادب ومايتعلق به ومما يدل على ذلك ما ذكره
في رسالته التي وجهها بدعوة الاطباء وكان ابن رضوان أعلم بالطب وأعلم بالعلوم الحكمية وما
يتعلق بها وكان ابن رضوان أسود اللون ولم يكن بالجيل الصورة ولهذه المقالة في ذلك ردينهما
على من عيره بقبح الخلقة وقد بين فيهما أنه ليس من شرط أن الطبيب الفاضل لا يجب أن يكون وجهه
جميلا وكان ابن بطلان أكثر تابعا في علي بن رضوان من هذا القبيل وأشباهه ولذلك يقول
فيه في الرسالة التي وجهها أبو الحسن الاطباء (الطويل)

فلما ابتدى لذة وابل وجوهه ٭ تكمن على أعقابهم من الندم
وثار وأخفين الكلام تسترا ٭ ألا ليتنا كذا ركاه في الرحم

وكان بلقب قماح الجن وسافر ابن بطلان من ديار مصر الى القسطنطينية وأقام بها سنة
وعرفت في زمنه أوبئة كثيرة (ونقلت) من خطه فيما ذكره من ذلك ما هذا مثاله قال
ومن مشاهير الاوبئة في زماننا الذي عرض عنده طلوع الكوكب الاثاري في الجوزاء من
سنة ست وأربعين وأربعمائة فان في ذلك السنة دفن في كنيسة لوقا بعد أن امتلأت جميع
المدافن التي في القسطنطينية أربعة عشر ألفا نسمة في الخريف فلما توسط الصيف
في سنة سبع وأربعين لموت النبل لماتت في القسطنطينية والشام أكثرأهلها وجميع الغرباء
الا من شاء الله وانتقل أو باءالى العراق فأتى على أكثرأهله واستولى على الخراب
بطروق المساكر المتعادية واتصل ذلك بها السنة أربع وخمسين وأربعمائة وعرض
للناس في أكثر البلاد ذرو حمودية وأورام الطحال وتغير ترتيب نوائب الحميات
واضطرب نظام الامجاري من اختلاف علم القضاء في تقدمة المعرفة وقال أيضا بعد ذلك
ولان هذا الكوكب الاثاري طلع في برج الجوزاء وهو طالع مصر أوثم أوباء في القسطنطينية
بتهمان النبل في وقت ظهوره في سنة خمس وأربعين وأربعمائة وصح انذار بطليموس
القائل أبو بل لأهل مصر اذا طلع أحد ذوات الذوائب وانجمهم في الجوزاء والمازل
زحل برج السرطان تكمل خراب العراق والموصل والجزيرة واختلت ديار بكر وديار
ومصر وفارس وكرمان وبلاد المغرب والبمن والقسطاط والشأم واضطربت أحوال
ملوك الارض وكثرت الحروب والغلاء والوباء وصح حكم بطلميوس في قوله ان زحل
والمريخ متى اقترنا في السرطان زلزل العالم (ونقلت) أيضا من خط ابن بطلان فيما ذكره
من الاوبئة العظيمة المعارضة للعلم بعقدها علماء في زمانه ذلك ما عرض في مدة بضع عشرة
سنة بوفاة الاجل المرتضى والشيخ أبي الحسن البصري والفقيه أبي الحسن القدوري
وأفضى القضاة الماوردي وابن الطيب الطبري على جماعتهم رضوان الله ومن أصحاب
علوم القدماء أبو علي بن الهيثم وأبو عبدالله بماهي وأبو علي بن السمح ومساعد الطبب

Fig. 1 A 'disaster' attributed to the Crab Nebula supernova of 1054. Ibn Abi Usaybia's medical textbook says that 'One of the well-known epidemics of our time is that which occurred when the spectacular star appeared in the year 446 AH. In the autumn of that year fourteen thousand people were buried in the cemetery of St Luke after all the other cemeteries of Constantinople had been filled. In midsummer, most people in old Cairo died; the epidemic spread to Iraq and affected most of the population.'

we do, and so it is possible to say quite precisely in which direction it lay, the asterism Thien-kuan, which includes the star Zeta Tauri. If astronomers look in this direction today they see a milky nebula, known as the Crab (Plate 13). Photographs show a milky glow, surrounded by a network of filaments; the whole nebula is reminiscent of the shapes of explosions in movies like *Star Wars*. In fact, this is what the Crab Nebula is—an explosion. Its present size, some 1000 years after the supernova of 1054 was first seen, is about 3.5 light years in radius, corresponding to an average expansion speed of 1300 km s^{-1}.

The mass of the filaments in the nebula is several times the mass of the Sun, and the kinetic energy of the explosion of the Crab some 10^{44} J. If we assume that the Crab and the supernova of 1054 are indeed related, and therefore that the supernova was at the same distance from the Earth as the Crab (5500 light years away), we can integrate the area under the light curve of the supernova as observed by oriental astronomers to obtain the total radiated energy from the supernova: this too is some 10^{44} J. We would expect a coincidence like this in the explosion under general equipartition principles: the fundamental energy release from the explosion can eventually take several forms which, if they have ever been in equilibrium one with another in the explosion process, will be equally distributed.

The outward flow of the explosion is something that it has been possible to perceive over the past few decades in which large scale photographs of the Crab have been obtained. The individual filaments are small enough that small displacements of the filaments can be seen from one photograph to another after only a decade's interval. It is possible to track the explosion back to its origin in 1054 or thereabouts.

The Crab Nebula therefore represents a coincidence in energy and time as well as in direction with the supernova of 1054, making astronomers certain that the nebula which they see today has its origin in the event seen by Anasazi Indians and oriental astrologers in the past.

Types of supernovae

The perceived phenomenon of a supernova is a bright flash lasting a length of time from a matter of months to a few years with an integrated electromagnetic radiative energy of some 10^{44} J, and arising from the onset of a cosmic explosion of about that energy, followed by the outward flow of material from the explosion on a timescale of millennia. Supernovae have more than one inherent cause; the first clues that this was so came from astronomical spectroscopy.

12. The supernova of 1054. Off the lower cups of the crescent moon is the image of a bright star, painted by Anasazi Indians on the wall of a cave at Chaco Canyon. Now high above the ground, after the fall of the platform from which the sunrise was observed to mark the seasons, and safe enough for birds to nest nearby, this depiction of the appearance of supernova 1054 in the dawn sky on 4th July, 1054 is 'signed' by a hand print.

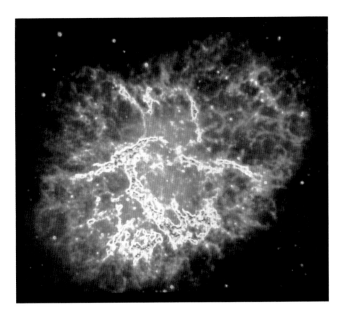

13. The Crab Nebula. The exploding body of the supernova of 1054 shows as outflowing filaments in this electronic picture assembled by Derek Jones with the Kapteyn Telescope on La Palma.

Astronomers categorize supernovae into two types, based on their optical spectra. Type I supernovae do not have hydrogen emission lines in their spectra while Type II supernovae do. In early work, it developed that Type II supernovae were only ever seen in the spiral arms of galaxies, places where there are recently formed stars (older stars have migrated away from their places of origin in the interstellar material in the spiral arms). From this developed the supposition that Type II supernovae are events in the lives of relatively short-lived stars. Massive stars are very much brighters than typical stars like the Sun, and although they have more fuel (in proportion to their mass), they radiate energy at a rate that is proportionally much greater. Supernovae of Type II seem therefore to be connected with explosions of massive stars (those more massive than 5 times the mass of the Sun). The origin of supernovae of Type I is not known, but it seems likely that these supernovae are explosions of white dwarf stars.

The clear division between Type I and II supernovae has recently been complicated by the recognition of Types Ia and Ib, with those in class Ib being related to Type IIs. As in other observational sciences, such as biology, the taxonomy of supernovae has started from superficial appearances but developed as astronomers gained deeper understanding of the underlying causes. The deeper understanding has however rendered confusing the original naming system.

In this article, I do not want to go any further into these arcane matters. I am going to restrict my subject matter to Type II and Type Ib supernovae, the explosions of massive stars. These are the most common types of supernovae.

Supernovae in space (distant) and time (infrequent)

Supernovae (of Types II and Ib) are the deaths of (massive) stars; their rate of occurrence is determined by the number of (massive) stars and their lifetime. There are large numbers of such stars in our Galaxy but they live a relatively long time, so stars die as supernovae in our Galaxy only about once per century. As it happens the last supernova known in our Galaxy was in 1604—nearly four centuries ago. Why has there been such a relatively long interval since the last galactic supernova? Astronomers put it it down to chance (there is a probability of the order of a few per cent that a gap this long would occur in a Poisson statistical series). Bad luck to astronomers.

The supernova of 1604 occurred essentially before the invention of scientific equipment, e.g. before the telescope. To know about close supernovae, astronomers study historical records, which, however fascinating, remain tantalizingly incomplete and confusing, or they study the expanding remnants of the explosions, which lie some centuries or millennia distant from the smoking gun.

Most knowledge of supernovae themselves, until recently, was from the study of supernovae in galaxies very distant from our own. Astronomers discover them at the rate of one or two per month. The study of the light from such distant supernovae is hampered by the diluting effects of distance. A supernova at 10 million light years distance is dimmed by a factor of a million compared with one at 10 000 light years—this is a veritable handicap.

This does not all add up to a favourable situation and astronomers deserve great credit for piecing together accurate stories about supernovae from events at great distances of space and time. Their calculations were shown to be remarkably accurate when a relatively nearby supernova exploded in 1987, in the galaxy next door if not in our own Galaxy.

SN1987A in the Large Magellanic Cloud

On 23 February 1987 the news reached Earth of a star exploding in the nearest galaxy to ours. This galaxy is called the Large Magellanic Cloud. It looks like, but is not, a cloud in the night sky. It was known to Australian aboriginals, South African Bantu, and Amerindians long before reports of it reached Europe from the southern explorations by Magellan. It is one of a pair, and is actually smaller than the Small Magellanic Cloud, which is longer, but thin and—because we see it end on—diminished by perspective. The name Large Magellanic Cloud thus leaves a lot to be desired—but LMC serves as a convenient label for what is a minor satellite galaxy to ours, a mere 170 000 light years away, bearing the same relation to our Galaxy as the Moon does to Earth.

The supernova was first discovered by accident by astronomer Ian Shelton on a photograph of the LMC taken with a telescope in Chile for routine purposes. As the first supernova discovered that year, it was designated SN1987A (Fig. 2). Over the succeeding months astronomers watched it grow so bright that it could be seen with the unaided eye, and equalled all the rest of the stars in its galaxy put together. With the benefit of modern equipment and large telescopes, both on the ground and in space, astronomers were able to measure the power output from the supernova, not just by eye in visible light, but over the broad electromagnetic spectrum—infra-red, light, ultraviolet, and X-radiation. The power output from the supernova faded from maximum, falling exponentially with a half-life of some 80 days.

Fig. 2 Supernova 1987A in the Large Magellanic Cloud. A pair of 'before' and 'after' pictures taken with the UK Schmidt Telescope in Australia shows a prominent nebula and numerous faint stars in our nearest neighbour galaxy and, lower right, the bright image of the exploding 'new' star, far outshining all the others.

SUPER!nova

What triggers a supernova? Where does its enormous energy come from?

A supernova is triggered by the collapse of a star's interior core. The release as heat of gravitational potential energy in the collapse of the central core is picked up by the star's outer layers, which are blown off, at speeds of up to $c/10$ (where c is the speed of light). At such a rate the star's surface grows to the size of the solar system in, say, a day. The rapid increase in surface area is the reason that the star brightens so spectacularly.

Why does the core collapse? To answer this we should first ask why stars mostly do not collapse.

A given layer in a star is in equilibrium as a balance between downward and upward forces. The downward forces are, first the weight of the layer itself, drawn towards the star's centre by its force of gravity, and second the weight of the layers of the star above, which press down on the given layer. The upward force is the pressure from below the layer, not only gas pressure arising from the temperature of the internal material, but also the pressure of radiation moving upwards, from the centre of the star where it is made to the surface where it is radiated.

The radiation which supports each layer of the star originates in nuclear fusion reactions in the star's core. Four hydrogen nuclei (protons) are fused to lighter helium nuclei with consequent release of energy in the form of γ radiation. There is a lot of hydrogen in a star so that, although the equivalent mass lost from a star in the form of energy is measured as millions of tonnes per second, the hydrogen fuel in the star lasts for a long time. Even so, the hydrogen fuel does eventually give out.

The star is able to adjust its internal structure as hydrogen is depleted. The core heats to higher temperatures so that three helium nuclei fuse to a carbon nucleus. There is not as much energy available in this fusion stage, nor in the subsequent stages which rapidly follow, so these stages only serve briefly to stave off the time at which all the sources of energy in the star's core run out. Thus while the hydrogen fusion stage may last 10^8 years and the helium fusion stage may last 10^4 years, the later stages of the fusion chain may be over in weeks or days. The upward radiation pressure dies away as the fuel gives out. There remain only downward weight forces and the result is inevitable—the core collapses.

The core's mass is about the mass of the Sun and its radius is measured in tens of thousands of kilometres. It implodes to a core of neutrons, a neutron star of radius some 10 km, at which state it constitutes a kind of atomic nucleus of mass number 10^{56}.

The gravitational binding energy of the neutron star is of order GM^2/R where R is its radius (10^4 m), M its mass (10^{30} kg) and G is the gravitational constant (7×10^{10} N m^2 kg^{-2}). This is the amount of potential energy released in the collapse, some 10^{46} J.

10^{46} is a big number and 10^{46} J is evidently a lot of energy. I could have made the number look more impressive by expressing the energy in ergs or electron-volts, because it is hard to grapple realistically with such big numbers. To appreciate the true size of this sudden energy release we need to get it into comparative perspective.

The mass-equivalent that has to be annihilated ($E = mc^2$) to produce 10^{46} J is getting on for 10 per cent of the mass of the Sun. Bear in mind that this comes from an object (the core) which is the mass of the Sun—in this sense the supernova energy conversion as a fraction of rest mass is one of the most efficient known, 10 times more efficient than the hydrogen nuclear fusion reactions in the core of the star before collapse.

There is clearly a discrepancy between the release of gravitational potential energy in the core collapse (10^{46} J), the observed kinetic energy of outflow of the expanding material in a supernova remnant (10^{44} J) and the electromagnetic energy which is radiated in the initial flash (the same). This means that over 95 per cent of the energy released by the collapse is dumped into an almost undetectable form, which does not equilibriate with the others. This has been known to astronomers from theoretical calculations, but was not proved until the supernova of 1987. Most of the energy of the supernova is radiated not as electromagnetic radiation but as non-interacting neutrinos.

Rate of collapse of the core and its power output

The gravitational potential energy of the collapsing core is mostly deposited in neutrinos. Neutrinos (literally 'little neutral particles') participate in weak nuclear reactions. They are generated in the interactions at high temperatures and densities between nuclear particles in stellar interiors, neutron stars, and the stages between, such as the collapse of stellar cores (i.e. supernovae). Neutrinos, unlike photons, generate practically zero pressure because, unlike photons, they do not readily interact with material. Once generated in the collapse of the core, neutrinos travel freely through the outer layers of the star, failing to support the star. The energy of the collapse is deposited, not only in neutrinos but also in nuclear synthesis—the build-up of large nuclides, whose pressure contribution in the star is also small.

Both these factors mean that, as the neutron star forms, the internal pressure above the neutron star's surface rapidly dies away. As a result, the collapse takes place at almost free fall.

In cartoons, when a character speeds off a cliff, he moves horizontally, stops in mid-air and then plummets straight down. We all know that it is more accurate to describe him as falling in a parabola, but the animator has separated the horizontal conservation of momentum and the vertical acceleration.

If I likewise separate the physics of the core collapse (which is actually complex and interdependent) into serial stages, then the gravitational potential energy of the core is converted to kinetic energy of implosion, and then into heat as the implosion is arrested as a neutron star is formed. Equating the gravitational potential energy to kinetic energy just before the neutron star is created (and ignoring factors of two, which have no place in this rush of approximations!), I can estimate the free-fall speed, V:

$$GM^2/R \sim MV^2.$$

I can thus estimate the free-fall timescale, t

$$t \sim R/V \sim (R^3/GM)^{1/2}.$$

This is less than a second.

The power output of the supernova is thus about 10^{46} J s^{-1}. The power output of the Sun is 10^{26} J s^{-1}, so the supernova has the power of 10^{20} Suns, about equal to the power of all the stars of the Universe (10^{10} stars in 10^{10} galaxies).

Production of neutrinos and their effects

If I continue with this cartoon-style physics, the energy of core collapse is deposited as heat in the neutron star. An energy of 10^{46} J in a body of mass 1 solar mass raises it to temperatures of 10^{10} K. The predominant cooling mechanism of bodies at this temperature is not by radiating photons (light, X-rays, γ-rays, and so on) but by radiating neutrinos. The neutrinos form a thermal spectrum at the surface of the neutron star, with energies about 1 MeV.

Neutrinos are famous, as Richard Feynman commented, for doing virtually nothing, just like your son-in-law. Their mean-free-path between interactions in material of density that of water (1 g cm^{-3}) is of the order of 1 light year. Such a column of material has a column density of 1 light year $\times$ 1 g cm^{-3} = 10^{18} g cm^{-2}. The column density through a star is about 10^{11} g cm^{-2} (the mean density of a star is about that of water, but its radius is about 1 light second, much less than 1 light year). The column density through a star is thus over a million times smaller than the column density in which neutrinos typically interact.

As a result, over 90 per cent of the energy released by the cooling neutron star moves quickly out of the star as non-interacting neutrinos. This is why nearly all the energy from the supernova is radiated in a near-undetectable form. This is why, impressive as it is when we see the light that a supernova radiates and the kinetic energy that is inherent in the outward flow, the energy release from a supernova is actually about 100 times more than we see.

However, the dense material near the cooling surface of the neutron star has a density up to around 10^{11} g cm^{-3}, and depth about 1 km, so its column density is approximately 10^{17} g cm^{-2}. This is sufficiently high that some neutrinos do interact with this dense layer. About 1 per cent of the neutrinos' energy is deposited in the envelope of material surrounding the collapsed core. In addition, material from the outer fringes of the core begins to rain down into the centre of the collapse and finds itself pounding on the hard surface of the neutron star: it bounces and the shock at the bounce propagates into the envelope of material outside.

The neutrinos and the shock-wave deposit energy into the region around the neutron star's surface, reverse the collapse of this material, and life off the star's envelope in an explosion.

The prompt burst of neutrinos from the supernova is very large (10^{46} J, divided into units of 1 MeV, implying that about 10^{57} neutrinos are created). Even when spread on to the surface of a sphere of radius 170 000 light years, their flux is of the order of 10^{14} per square metre. This was the flux of neutrinos at the Earth from SN1987A, which reached us at 07:35 GMT on 23 February 1987. Roughly this number of neutrinos (10^{14}) would have irradiated everyone on this planet (assuming that a person's cross section is about 1 square metre, depending mostly on his or her orientation relative to the LMC at that time).

Detection of neutrinos from SN1987A

Consider now a cubical tank of water, with each side measuring some 10 metres, such as might fill a large laboratory. Since its cross section is 100 m^2, 10^{16} neutrinos from the supernova of 1987 passed through it. The depth of water in the tank is 10^{-15} of a light year, that is 10^{-15} of the mean free-path of neutrinos in water. Thus about 10 neutrinos are likely to have interacted with the water in the tank. Two such tanks were operating as neutrino observatories at the right moment in 1987, one in Gifu, Japan and the other on the shores of Lake Erie, Ohio (Figs 3–5). Together they observed a dozen or so neutrinos in the wave that passed through the Earth that day.

The two neutrino observatories were in fact built and operated to watch for any decays of protons in the tanks of water. Although expectant

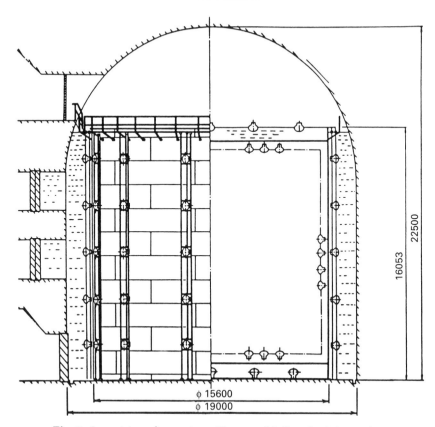

Fig. 3 A neutrino observatory. Photomultipliers look inwards to a 15 metre cubic tank of purified water, shielded by blocks from incoming radiation from radioactive decays and guarded by outward-looking photomultipliers, all underground in a mine to reduce cosmic rays.

physicists have watched many Avogadro's numbers of protons in the tanks for many years, no such decays have ever been observed. This shows that protons have life-times in excess of 10^{34} years (this is of interest in discriminating amongst various versions of Grand Unified Theory). It had been recognized that the experimental design of these proton decay experiments would detect neutrinos from a supernova in or near to our Galaxy, though it had also been recognized that the probability that such an event would occur while the experiments were active was low (of the order of a few per cent per year of operation). No doubt the happy observation of the LMC neutrinos has raised the morale of the physicists operating the proton decay experiments.

Neutrinos of energy 1 MeV (typical of those from the supernova) interact with water in two ways. There are three different kinds (at least) of

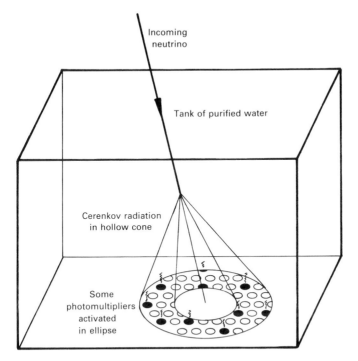

Incoming
neutrino

Tank of purified water

Cerenkov radiation
in hollow cone

Some
photomultipliers
activated
in ellipse

Fig. 4 Detection of neutrinos. If an incoming neutrino interacts
in a tank of water it produces a relativistic electron or positron,
which radiates Čerenkov light in a cone. Photomultipliers in the
walls of the tank detect the light which forms a characteristic
elliptical pattern (conic section).

neutrinos, the neutrinos associated with electrons, muons and tauons
respectively. To each kind of neutrino there corresponds an antineutrino.
Any kind of neutrino (ν) or antineutrino ($\bar{\nu}$) can scatter off an electron (e)
which is orbiting a water molecule in the tanks:

$$\nu + e \rightarrow \nu + e.$$

The incoming neutrino has energy 1 MeV and the electron picks up a
fraction of this energy. It is broken free from the water molecule to which
it is bound by only a few eV. The electron is thus relativistic, with a speed
near to c. Since the electron is light and has almost zero binding energy,
the scattered electron 'remembers' the momentum of the inbound neutrino,
and points away from the LMC.

Additionally to scattering on an electron, an electron antineutrino ($\bar{\nu}_e$)
can interact with a free proton (p) to make a neutron (n) and a positron (β^+).
Protons exist in water as the hydrogen nuclei in water molecules:

$$\bar{\nu}_e + p \rightarrow n + \beta^+.$$

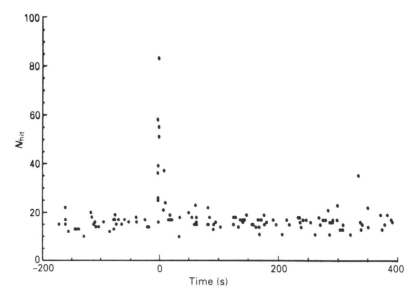

Fig. 5 Seven neutrinos from SN1987A. The number of photomultipliers active at a given moment in the Kamiokande Observatory in Japan is usually 10 to 20, triggered by noise events. But,
in a unique event at 07.35 GMT on 1987 Feb 23, seven events
triggered up to 85 photomultipliers as the observatory recorded
Čerenkov radiation. The seven neutrinos were attributed to
SN1987A because it was not credible that such a unique, otherwise inexplicable event (seen simultaneously at another similar
observatory in Ohio) should occur by chance at the time of a
nearby supernova.

As in any typical nuclear reaction, the positron picks up an energy of order
1 MeV: it too is relativistic. Because the proton is heavy, the positron's
momentum vector is not readily relatable to the incoming direction of the
neutrino.

In both interactions a neutrino yields a charged particle travelling
through water with speed near c. The refractive index of water is 4/3, so
the speed of light in water is $3c/4$. When a charged particle moves through
a medium faster than the speed of light in the medium, the result is
Čerenkov radiation, photons radiated from the kinetic energy of the
charged particle.

A neutrino observatory thus consists of a tank of water surrounded by
photomultipliers looking inwards; neutrino interactions are recognized
by the burst of light which they cause. Since Čerenkov radiation is emitted in a cone centred on the flight path of the charged particle, the
photomultipliers in the wall of the tank are fired in a pattern that forms a
conic section (ellipse or hyperbola). Sophisticated pattern-recognition

techniques can discriminate between neutrino interactions and other random events. Random events were minimized in the experimental design by using water purified of radioactive salts and by locating the tank below ground to reduce the cosmic ray background.

The neutrino bursts detected simultaneously on opposite sides of the world on 23 February 1987 cannot have been due to local events—a single cosmic (at least global) event must have been the cause. It is not credible that the only such event of this type observed would be by coincidence located at the beginning of the nearest supernova event for 400 years. This is the basis for connecting the neutrino burst with SN1987A.

Each neutrino burst was about one second in duration. This confirms the timescale for the formation of the neutron star in SN1987A.

Neutrino properties from the SN1987A neutrino burst

The neutrinos had been travelling at speeds near the speed of light, over a distance of 170 000 light years. Unfortunately the start of the outburst of light from the supernova was defined only with an accuracy of hours by the spacing of chance observations of the LMC showing it without and with the supernova. Anyway, there is a significant delay of many tens of minutes between the collapse of the core of a star (when the neutrinos are created) and the star's expansion and brightening (when it begins to make itself known as a supernova). This means that it is possible to relate the neutrino speed to the speed of light only to an accuracy of an hour or two in 170 000 years (10^{13} seconds), or one part in 10^9.

However, if the neutrinos travelled at different speeds (*dispersion*) the neutrino pulse would have broadened. The upper limit of about one second to the dispersion after 170 000 years puts a stringent limit on the range of speeds of the neutrinos (they travelled at the same speed to within one part in 10^{13}).

The lack of dispersion implies that neutrinos do indeed travel at a single speed to a very high accuracy, and we know that this speed is very near to the speed of light. The only kinds of particles which travel exactly at the speed of light are ones whose rest masses are zero, so the rest masses of neutrinos must be small. It has long been known that electron neutrinos have undetectable rest mass (less than 18 eV energy equivalent), and the standard theory is that their rest mass is zero, like that of photons. If so, they travel at the speed of light exactly, without any dispersion in their speeds. The experimental upper limits on the rest mass of muon and tauon neutrinos are weaker than that for electron neutrons (less than 0.25 MeV and less than 70 MeV respectively). The lack of dispersion of the supernova neutrino burst, presumably a mixture of all kinds of neutrinos, implies an upper limit to the rest mass of all neutrinos of about 10 eV, which is comparable with the laboratory upper limit to the rest mass of electron neutrinos.

The energies of the detected neutrinos imply an energy flux in the neutrinos which is directly related to the binding energy of the neutron star which was formed. There is excellent agreement.

In a thermal neutrino spectrum (and if all neutrinos have zero rest mass) there are equal numbers of electron neutrinos, muon neutrinos, and tauon neutrinos, and of the equivalent antineutrinos (each kind numbers one sixth of the total). The more kinds of neutrinos which are created, the lower the energy which lies in each, since a given amount of energy is spread wider. The properties of the SN1987A neutrino pulse, related to the binding energy of a neutron star, are consistent with there being three species of neutrino (plus the corresponding antineutrino species). The supernova neutrinos confirm the standard theory in this respect, as do the most recent particle physics experiments.

The bottom line for the detection of neutrinos from SN1987A was that this was a scientific triumph, with observations marvellously in accord with astronomical prediction of the collapse of stellar core to form a neutron star and the tentative standard theory of neutrinos.

The key component to the theory of neutrino interactions is the neutral current theory mediated by the Z^0 particle which was discovered at CERN only a few years before the supernova was seen in the LMC. If the LMC was, say, 10 light years closer than 170 000 light years, the wave of neutrinos would have passed the Earth in 1977, before the experiments had been set up and before physics was equipped with the appropriate understanding to take advantage of them.

Nuclear synthesis in supernovae

Temperatures of 10^{10} K, and densities of 10^{14} g cm^{-3} imply nuclear reactions. A considerable fraction of the available energy of the core collapse thus finds its way into the creation of new elements (nuclear synthesis) from the simple fusion products of the core material produced in the earlier life of the star. Predominant nuclides of gold and uranium are just two of the species predominantly made in this way.

It is intriguing to look at a piece of gold and imagine its history. Five billion years ago or more, in the early times of our Galaxy, unknown supernovae synthesized gold, enriching the interstellar material nearby, from which condensed the Sun; the detritus from this process, in orbit around the Sun, produced the planets, including the Earth, from which we now dig supernova material to make gold teeth and jewellery.

Proof of nucleosynthesis in SN1987A came from the observation of three signatures of radioactive isotopes of nickel and cobalt produced on 23 February 1987. The small lifetime of the isotopes observed guarantees that they are related to the supernova event, not to any previous occurrence.

The radioactive isotope produced in greatest amounts in SN1987A (0.07 times the mass of the Sun) was ^{56}Ni. ^{56}Ni (which has a half life of 6 days) quickly decays to ^{56}Co (half life of 77 days). The energy (energetic electrons and gamma rays) released by the radioactive decay of these isotopes, which lie at the centre of the supernova in the nuclear synthesis region, is absorbed by the surrounding, expanding envelope of the supernova and re-radiated as heat and light. This source of energy keeps the supernova shining longer than its natural fading time; the supernova flash is prolonged by this mechanism. The radioactivity is like a spring, wound in the explosion, which runs down and supplies energy into the supernova at a later time. The spring runs down at the radioactive decay rate. ^{56}Co has a half-life of 77 days and it was ^{56}Co which was responsible for the exponential decay (half-life of about 80 days) of the electromagnetic radiation from SN1987A throughout 1988–89 (Fig. 6).

^{56}Co decays to ^{56}Fe with the emission of 847 keV γ-rays. As the

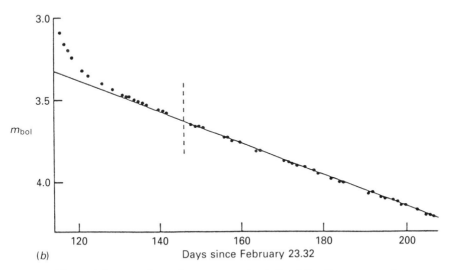

(b)

Fig. 6 After 140 days, the power output, P, of the Supernova of 1987 settled to an exponential decay, as shown in this logarithmic plot ($m_{bol} = 2.5 \log P$). The time scale of the decay was identical to the timescale of the radioactive decay of cobalt-56, which had been created by nuclear genesis in the explosion and which was the dominant input of power into the material of the supernova from this time.

supernova expanded, it became possible to see through holes in the expanding material to its central regions where the ^{56}Co lay. The γ-rays were detected by a satellite, Solar Max, that had been placed in orbit in 1981 to study the Sun and that was still operational in 1987 (Fig. 7). The 847 keV γ-rays penetrated the body of the satellite and activated its detectors even though the satellite was pointed to the Sun, at right angles

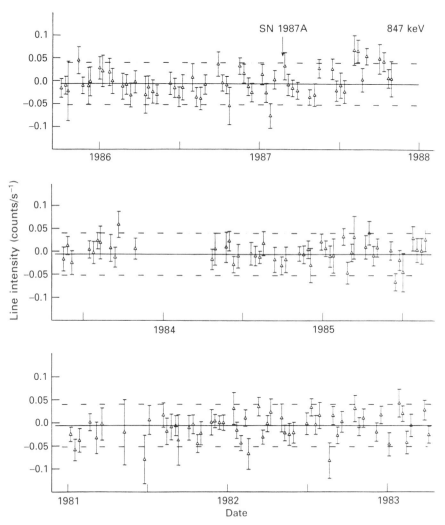

Fig. 7 The signal from gamma rays of 847 keV energies from the Large Magellanic Cloud, as detected by Solar Max. No signal was detected throughout the first six years of the satellite's life, but, lo, in the seventh year, a positive signal appears six months after the supernova, after the exploding debris of the star has cleared enough to let out gamma rays.

to the direction of the LMC. It would have been difficult to distinguish γ-rays from the supernova and γ-rays from other sources (including spurious noise events) if it had not been for the motion of the satellite around the Earth. At times when the Earth lay between the satellite and the LMC, its bulk extinguished the satellite's view of the LMC. It was possible to use the body of the Earth as a shutter during the orbit of the satellite to chop the signal from the LMC and detect the flux of 847 keV γ-rays above background. Infra-red spectra of SN1987A obtained into the second year of the supernova with the Anglo-Australian Telescope also show emission lines of cobalt reducing in intensity while emission lines of iron increase, as ^{56}Co decays to ^{56}Fe, at the correct exponential rate.

^{56}Co was not the only radioactive cobalt isotope which was produced in the supernova explosion. By the end of the second year, the intensity of the cobalt emission lines in infra-red spectra had reached a quasi-plateau of emission from another cobalt isotope. (Spectral emission lines of cobalt depend on the electron structure of cobalt, rather than its nuclear properties, and do not show what isotope lies in the centre of the emitting atom.) After two years essentially all the ^{56}Co had decayed, and the cobalt emission lines were from the longer-lived ^{57}Co isotope. γ-rays from ^{57}Co decay have recently (1992) been detected by the Compton Gamma Ray Observatory.

As the longer-lived radioactive isotope of ^{57}Co, and then later in 1992 the even longer-lived isotope ^{44}Ti (with a half-life of 77 years) became the dominant continuing supply of energy into the envelope of the supernova, the light-curve of the supernova flattened out to the correspondingly slower exponential decay.

Those observations startlingly confirm the proposal that nuclear synthesis of heavy elements from the simpler elements of the Big Bang takes place in stars and that without supernovae we would be without some naturally occurring isotopes and with much reduced quantities of some elements (like gold).

SN1987A in 1993 and the future

When it first appeared, the supernova was point-like, as are all stars at the distance of the LMC. Now, several years afterwards, the supernova shows perceptible structure when viewed in ground-based telescopes and very clearly shows structure in images taken with the Hubble Space Telescope (HST), with its higher resolution in an environment free from the blurring effects of the Earth's atmosphere. HST images show the supernova's growth as the explosion proceeds (Fig. 8).

Eventually, over centuries, the supernova will grow to appear like the

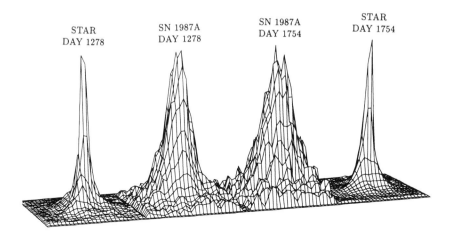

STAR
DAY 1278

SN 1987A
DAY 1278

SN 1987A
DAY 1754

STAR
DAY 1754

Fig. 8 SN1987A from HST. The clear view of the Hubble Space Telescope, orbiting above the blur of the Earth's atmosphere, has recorded the growth of the explosion of SN1987A over 15 months. Two images of SN1987A (centre) are clearly broader than the images of point-like stars (left and right), and the later image of SN1987A is broader than the earlier one, representing growth of the expanding nebula at a rate of 5000 km s^{-1}.

Crab Nebula. Over millennia or hundreds of millennia, SN1987A will disperse into the interstellar medium of the LMC. It can be expected to contribute to the formation of future stars and planetary systems. Perhaps civilizations will evolve on these planets to keep the calendar, to make gold jewellery, to study astronomy, to create scientific societies like the Royal Institution, and to marvel at and study supernovae. In the future, as now and in the past, I expect supernovae to continue to contribute to the production of extraordinary devices (called 'people') as a way of learning about themselves!

Bibliography

This article was based on material in the following books:

Murdin, P. (1989). *End in fire: the supernova in the Large Magellanic Cloud.* Cambridge University Press.

Murdin, P. and Murdin, L. (1984). *Supernovae.* Cambridge University Press.

PAUL MURDIN, O.B.E., Ph.D.

Born 1942, he was educated at Oxford University and the University of Rochester NY. Held posts at the Royal Greenwich Observatory, the Anglo-Australian Observatory and the Royal Observatory Edinburgh. In 1971 co-discovered the black hole in Cygnus X-1, the first black hole identified. In 1976 co-discovered the optical emission from the Vela pulsar, the second pulsar firmly identified with a supernova. From 1981 to 1987 set up the UK-Netherlands telescopes on La Palma in the Canary Islands. From 1991 has been Director of the Royal Observatory Edinburgh and in charge of the two British telescopes on Mauna Kea in Hawaii. Author of ten popular books on and broadcaster on astronomy.

.

Tunnelling under the Channel

JOHN V. BARTLETT, CBE, MA, F. Eng., FICE

Spike Milligan wrote an autobiography entitled 'Hitler; my part in his downfall'. I am not quite sufficiently immodest to confess that this discourse should have the title 'The Channel Tunnel; my part in its achievement', but I can say that, like Spike, I gave it some of the best years of my life. My point is that I am only one of many engineers who have contributed to the design and construction of the Channel Tunnel.

I graduated in the early 1950s in Mechanical Sciences and then in Law. I was eating dinners in Gray's Inn towards being called to the Bar, when I joined the well-known contractors John Mowlem & Co. as a trainee engineer, and promptly found myself driving tunnels under Gray's Inn. I soon decided to become a full-time civil engineer, and have never regretted it.

I was lucky to be in the right place at the right time: there was a rapidly increasing demand for tunnels all over the world. The urbanization of Europe in the 19th century which led, for instance, to the construction of numerous tunnels under London, had become a world-wide phenomenon after the Second World War. There are now about 200 cities with a population of a million or more, all of them needing sewers and scarcely one of them not planning or building an underground railway.

For some years after the Second World War, tunnelling even in Europe and North America was a somewhat primitive exercise carried out by craftsmen and heavy labour. Great physical strength was needed and the normal work pattern was 12-hour shifts, for six days one week and six nights the next. The leading miners had special skills, particularly in supporting the exposed faces of ground, generally using timber.

The coal mining industry was large enough to justify the research and development costs of extensive mechanization, but civil engineering tunnels for sewers, underground railways, etc. tended to be one-off problems in much more variable geology and not on a sufficiently large scale to warrant the development of expensive machinery.

But there were circumstances where mechanical tunnelling machines could be used successfully. In London, especially north of the Thames, there is a stratum of relatively uniform London blue clay: a material hard and strong enough to stand unsupported for many hours, yet soft enough to be excavated by steel tools. It is impermeable, so any tunnel driven through the London clay without penetrating the aquifers above or below the clay remains dry. When the electric motor became capable of driving a passenger train in the 1880s, the 'tubes' under London became feasible and tunnelling machines (known as digger shields) were developed: primitive by today's standards but faster than tunnelling with hand-held tools.

Another place where mechanical tunnelling appeared feasible was under the Channel between England and France, where the geology indicated a continuous stratum of chalk all the way across Channel and, in 1881–82, tunnelling machines designed by Colonel Beaumont successfully drove a small circular tunnel near Folkestone about 2 km in length out to sea, and performed similarly near Calais. His machines were pneumatically driven and featured a rotating cutting head like all the most modern machines dealt with in this article. But that early attempt, like others since, was stopped by the British government.

To know which aspects of tunnelling under the Channel to cover in this brief article is difficult. When I first asked my firm for background information for this article on the current construction work, they sent me over 20 professional papers dealing with different aspects! Most readers will have some recent information about the present project now approaching completion, so it would be wrong for me to give a fulsome overall impression of the project. Instead, I will concentrate on the following aspects of which I have some personal knowledge: the geological investigations and assessments, the development of the tunnel boring machines, and the design of the tunnels they helped to build.

The arrangement of having a service tunnel (driven as a pilot tunnel) between two running tunnels (the tunnels in which the trains run), was established by my old firm, Mott, Hay, & Anderson, in the 1930s (Fig. 1), and this forms the basis of the whole project. As well as carrying essential services and providing an evacuation route, the service tunnel acts as the main ventilation duct. Ventilating the Channel Tunnel, thanks to the electric traction, is not a major problem: the air in the service tunnel will be kept at slightly higher than atmospheric pressure, providing controlled forced ventilation of fresh air into the running tunnels. That also means that in case of fire in a running tunnel, the service tunnel remains a

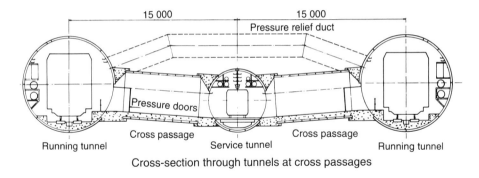

Cross-section through tunnels at cross passages

Fig. 1 A cross-section of the service tunnel and running tunnels.

smoke-free haven. Airtight doors are provided at all connecting cross passages.

'Piston relief ducts' have been constructed between the two running tunnels to obviate the need for trains to push a column of air ahead of them, and to reduce the vacuum behind trains. These measures reduce the power requirements for locomotives by hundreds of horsepower. The need for aerodynamic efficiency is quite exceptional and was a big factor in the design of tunnel linings with a smooth interior surface, and in deciding to have a continuous concrete railway track bed.

Why do we need a Channel Tunnel? Surely ferries are more flexible and modern roll-on roll-off ferry and aircraft services can be expanded to meet future demand? Would not the large and, to some extent, speculative investment in the Tunnel find a better home elsewhere? In my opinion the most important single reason why we need the tunnel is that continental Europe is completing a vast modern rail network, and unless Britain becomes a part of it, major disadvantages will result. Unfortunately, viewed from European railways, the Channel Tunnel at present looks like a short and minor spur line terminating near Folkestone. Surely the right answer is for us to rebuild our basic network to the same standards as the rest of Europe?

In addition to these strategic factors, there are immediate advantages which are different but substantial, and are represented by the three services which will operate through the completed tunnels:

1. The drive-on drive-off shuttle service will offer a quicker and more reliable service to all road traffic between the English and French terminals. It will be quicker not only because shuttle trains will be faster than ships, but also because they will be frequent and

because booking ahead will not be necessary. It will be more reliable because there will be no delays caused by bad weather.

2. Express passenger trains will operate. Anyone who has made frequent business trips from London to Brussels or Paris (as many of those involved in the Channel Tunnel have had to) will have suffered surprisingly numerous disruptions and delays. Even on our own antique railway system, Intercity has shown that trains can offer a faster service than aircraft over distances of 300 km and more. Express passenger trains from Paris and Brussels to London will be a substantial improvement on existing arrangements.

3. Freight trains running through the tunnels to and from destinations all over Europe will carry many thousands of tons of goods (which would otherwise have to travel by road) more cheaply and reliably, catering for the enormous and continuing increase in our trade with continental Europe.

The incidental advantages are numerous, ranging from fewer night-flights, to reduced danger of ship collisions in the Straits of Dover where there are about 600 major ship movements per day and the ferries have to cross the main traffic lanes.

Tunnel routes and construction

The first offshore geological investigations were carried out in the mid-19th century by a French engineer, Thomé de Gaumond, diving with weights on his feet, and bladders of air to bring him back to the surface with samples of the bottom. Various tunnel routes have been investigated since then, but only the three attempts starting in 1964, 1972, and 1986 are important to the present scheme. Taking positional and bathymetric surveys for granted, the two means of investigation were boreholes and geophysical seismic surveys. Boreholes were carried out either from jack-up platforms or from dynamically positioned drill-ships. Jack-up platforms have been used for many years for river and offshore civil engineering work and, of course, increasingly for offshore oil and gas fields. The maximum depth of water near the tunnel is about 60 metres; jack-up platforms have been built capable of much greater depths but they are very expensive to hire and move around.

The congested state of shipping in the Straits of Dover, the busiest

14. Completed section of running tunnel (Channel Tunnel).

15. Photograph of part of the clean-up fleet in Prince William Sound.

16. Washing oiled beaches in Prince William sound with pressurized water.

waterway in the world, is one of the reasons for transferring traffic from the ferries to the rail tunnels. Despite the very comprehensive navigational controls through the Straits, the sinking of a series of boreholes in the navigation channels was a daunting task. There was a school of thought that a drill-ship with automatic dynamic positioning would be better than a jack-up platform for a series of boreholes in crowded waters. The principles of dynamic positioning are to equip a ship (or a floating drill-rig) with a number of thrusters, each capable of producing a controlled thrust in any horizontal direction; reference beacons are placed on the sea-bed in desired positions close to the borehole, and are linked to the thrusters by computer controls which produce the requisite thrusts to keep the ship in the correct position relative to the beacons, automatically nullifying the effects of wind, wave, tide, and eddy.

If that can be achieved, then a well-managed drill-ship can move more swiftly from one borehole site to the next, and be less of a navigational obstruction to other shipping for a shorter time. However, when such a drill-ship was put to work on the tunnel investigations in 1972–73, bad weather was more of a problem than anticipated, and the dynamic positioning equipment was less than perfectly reliable.

Jack-up platforms performed satisfactorily and marine boreholes were made as follows: 75 in 1964–65, 21 in 1972–73, and 19 in 1986 (a total of 115).

The early boreholes in Colonel Beaumont's time had established that the simplified geology (from the surface down) consisted of:

(1) superficial deposits (often absent on the sea-bed);

(2) Middle Chalk, which is white, weathered to a variable depth, and fissured, and therefore wet;

(3) Lower Chalk, consisting of white chalk near the top, but with the clay content increasing with depth, becoming a virtually impermeable, grey chalk marl, which was identified as the best stratum through which to tunnel;

(4) the Gault clay, also impermeable but weaker than the chalk marl and more liable to expand into any open excavation and put an additional strain on to tunnel linings. The Gault clay stratum is about 45 metres thick on the English side of the Channel;

(5) Lower Greensand—a mixture of sands and clay down into which none of the Channel Tunnel structures have penetrated.

The 1964–65 campaign, with 75 marine boreholes, and some early geophysical surveys, greatly improved our knowledge of the profile between England and France, but in the course of the work, a valley filled with alluvium was found in mid-channel, now known as the Fosse Dangeard (Fig. 2). The alluvium extends to a depth of 80 metres below the sea-bed, well below the level of the tunnels, and its presence amply justifies the need for thorough site investigations, because if a tunnel had run into the Fosse unexpectedly, the result might have been calamitous.

The 1972–73 and 1986–68 campaigns were of course able to build upon the earlier work. By 1986, much more was known about the engineering properties of the various strata, thanks to instrumentation of the lengths of tunnel driven in 1974 before cancellation of the project. Also, by 1986, geophysical surveys had been greatly improved in quality, as had the various techniques of 'downhole investigation'.

The improvements in the geophysical work, largely as a result of oil and gas exploration in the North Sea, were due principally to better electronic instrumentation, higher energy single pulse signals, and especially the multiple recording of reflections by strings of geophones towed behind the survey vessel. Digital tape recordings gave magnetic data that could be processed by 'computer stacking' and 'image enhancing' techniques, so

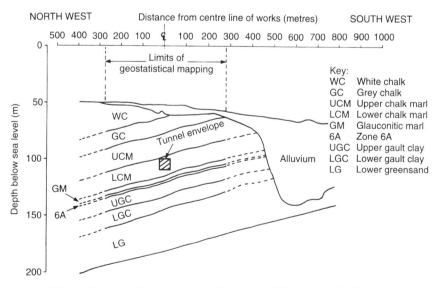

Fig. 2 Stratigraphy near Fosse Dangeard. WC, white chalk; GC, grey chalk; UCM, upper chalk man; LCM, lower chalk man; GM, glauconitic man; GA, zone GA; UGC, upper Gault clay; LGC, lower Gault clay; LG, lower Greensand.

just as the classic jazz records of the 1920s can now be reproduced without the hiss and background noise which some of us wrongly imagined to be an essential part of the ambience, so now the false echoes of the geophysical surveys can be eliminated and an enhanced image can be projected. I find it endearing that computer techniques are so often first used for games and entertainment and only later developed for more serious application.

By 1986, the start of the present project, continuous geophysical cover of a corridor 2 km wide with numerous transverse profiles, between England and France, had been correlated with a dozen new control bore-holes, giving quite clear indications of the top (and bottom) of the Gault clay, and therefore of the optimum tunnel levels across the Channel, plus or minus about 5 metres.

How should the exact route of the tunnels be decided? To continue my theme about computer games, imagine yourselves seated in an amusement arcade, playing a game of 'Space Invaders', except that you are going to invade France. You set off to conquer the Continent, leaving the English terminal at Cheriton, and passing under Shakespeare Cliff in the chalk marl (incidentally confirming Colonel Beaumont's engineering judgement of a century ago by intersecting his tunnel) and out to sea, taking care to keep at least 20 metres of chalk between you and the sea-bed. The first things you must avoid are 115 marine boreholes, all of them put down at great cost to assist you in choosing your route. If you hit any one of them, and it turns out not to have been conscientiously backfilled and grouted, then an ever increasing flow of water into the tunnel will result. But how accurately were they positioned? The maximum possible error for each borehole was derived, and found to be in the range 10–30 metres, depend-ing on the date and method used, so steer your underground space ship accordingly. But there are other limitations to your steering ability. This high speed railway must not negotiate curves of less than 4200 metres in radius as an absolute minimum, nor should gradients exceed 1.1% (one in 91). Luckily, on the English side, the angles of dip and the trend of the strata allow virtually all the undersea tunnels to be built in the chalk marl.

Now head for France, but not too far south or you will hit the Fosse Dangeard already mentioned and come to a sticky end. But wait, what is this approaching from the East? A French underground space ship, which set off from the French shore to invade England, has had a far bumpier ride. On the French side of the Channel, there has been much more flexing and folding of the strata, causing several faults and extensive fissuring. The dip of the strata is much more variable, being locally as much as 20°. The French space ship cannot manoeuvre to avoid all these features as it has to obey the strict rules of the high speed railway geometry, so it is unable to keep within the chalk marl and has a wet and bumpy ride for the first few kilometres, but the computer game has helped to minimize the difficulties and the two space ships meet. In fact, various suites of geological and alignment computer programs allowed a progressive optimization of the positions of the tunnels before and during their construction. Here was an ideal application of computer technology, allowing calculations to be varied in accordance with the latest information and interpretation in a way which, not long ago, would have taken so many man-years of calculations that the refinement of alignment would not have been possible.

However, engineering judgement was required as much as ever. For instance, once the British tunnels got clear of the cliffs, they hit an area where the chalk marl was brittle enough to be 'blocky'. Minor fissures let in some water and blocks of marl fell out of the top in positions where they had previously stood up. None of these minor variations had shown up in the previous tunnelling, or in the boreholes, or in the seismic surveys. The decision was taken to lower the tunnels, on the grounds that although that brought them closer to the Gault clay, the position of that interface was now known more accurately, and the tunnels would be in chalk marl of higher clay content. The decision seems to have been a good one, but will we ever be sure that it was? Possibly, by driving another tunnel in the future.

One of the many investigations undertaken was into possible danger from methane. Methane (otherwise known as firedamp) has often been a source of danger in underground workings, especially coal-mines, and concentrations of more than 5 per cent of methane in air form an explosive mixture. In view of the existence of the extensive Kent coal-fields, and remembering that the English worksite below Shakespeare Cliff was originally used as a coal-mine in the nineteenth century, careful investigations were made as to the possibility of methane seeping up into the Channel Tunnel works.

Attempts were made to trace methane in the Lower Greensand stratum in some of the exploratory boreholes, without success. When the Beaumont tunnel was pumped out in 1974, no methane was detectable. However, when one section of the Beaumont tunnel was again dewatered in 1988, methane was detected in the air up to a maximum of 6 per cent, i.e. within the explosive limits. After ventilation, the concentration in that tunnel and in all other tunnels never rose much above the minimum detectable concentration of about three parts per million. That isolated concentration in the Beaumont tunnel was thought to be caused by the presence of old creosoted timber.

During the years of tunnel construction, methane sensors were in operation at intervals throughout the workings, especially at sensitive points, including the tunnel boring machines (TBMs) and underground pumping stations, all linked to a control and communications centre on the surface, but with local as well as central alarm systems. The methane monitoring was part of a comprehensive environmental monitoring system which included other gases, temperature, humidity, atmospheric pressure, and air velocity.

The actions in response to methane concentrations (which were only ever activated by the isolated reading in the Beaumont tunnel) were as follows: 0.25 per cent, first alarm; 1.25 per cent, non-essential personnel withdrawn; 2.00 per cent, all personnel withdrawn; 5.30 per cent, lower explosive limit.

Needless to say, similarly comprehensive environmental monitoring will continue for the coming decades of public use of the tunnels, although the methane risk is surely very remote.

Tunnelling machines

Now I am going to blow my own trumpet as to my role in developing tunnel boring machines, or TBMs.

In the 1960s, British tunnellers were heavily and successfully engaged in building the Victoria Line, of the London Underground (Fig. 3), and

Fig. 3 Victoria Line digger shield (1963).

already now a whole generation of Britons have forgotten how difficult it once was to travel from King's Cross to Victoria, never mind from Walthamstow to Brixton! Almost none of them know that they pass almost directly under the Royal Institution as they travel.

The extensive use of digger shields and segmental linings expanded against the London clay was very successful. For British engineers, the timing was favourable: a highly automated underground railway was completed and operating successfully just when a substantial number of cities were realizing the need for such facilities, and tunnellers like myself were much in demand. A high proportion of large cities are, like London, situated at the lowest crossing point of a river, on alluvium, but instead of London clay, many of them are on saturated silt, sands, and gravels. Tunnelling through such ground used to be extremely difficult, requiring that the soil was chemically grouted, or frozen, or that the tunnels were driven 'in compressed air', that is all the tunnel workings were pressurized during construction to keep out water and flowing soil. All three measures were very expensive and time-consuming, and the use of compressed air results in men suffering, like divers, from compressed air sicknesses such as the bends, paralysis, and bone necrosis.

Bentonite is a clay which in water forms a thixotropic suspension, that is to say, it is a liquid as long as it is agitated, but it 'gels' when left undisturbed. Fuller's earth, found in Britain, is a sort of bentonite. Trenches in sand and gravel, if kept full of bentonite slurry, require no shoring. The bentonite penetrates the interstices of the ground and then forms an impermeable 'gel coat' which is itself supported by the hydrostatic pressure of the slurry in the trench.

Why not keep the working face of a digger shield full of a bentonite slurry at sufficient pressure to counter-balance any ground water pressure and support the working face? (Fig. 4). The slurry must be circulated and the spoil separated from it. A seal would be necessary to prevent the slurry (or the ground water) from leaking into the tunnel around the tail of the machine. Arrangements must be made to deal with any large stones which might be encountered to prevent their blocking any of the pipes.

The beauty of this method of supporting the working face is that by keeping the slurry pressure slightly higher than the ground water pressure, the whole face is provided with a positive but not excessive support pressure. This is a major improvement on using compressed air which inevitably provides too little pressure at the bottom of the tunnel, or too much at the top, or both.

That, then, was my invention; it became known as the bentonite shield, or slurry shield, and from it a whole family of TBMs have been developed that are capable of being used in the 'closed' mode, that is with the

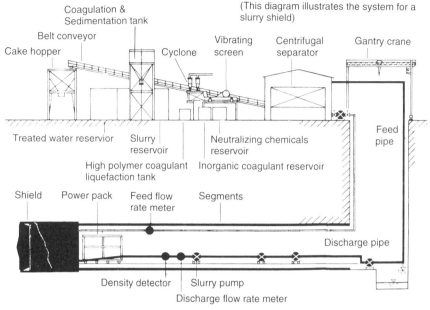

Fig. 4 Bentonite shield (1967).

working chamber pressurized. Germany and Japan, with large construction programmes, have been particularly active in developing these machines with several variations, including the earth pressure balance (EPB) machines (Fig. 5) which dispense with the bentonite slurry but, instead, mix the ground into a toothpaste-like consistency (with added water if necessary) and support the face with pressure maintained by allowing only a controlled quantity of the 'toothpaste' to be removed from the working chamber.

Several hundred such machines have now been built and used, and although the use of compressed air for some tunnels remains essential, the exposure of men to compressed air has been very greatly reduced. It gives me a warm glow to feel that my invention has helped to reduce the incidence of compressed air sickness in several thousand tunnellers.

Of particular importance for all these machines was the development of flexible seals to prevent the water or slurry from escaping into the tunnel; they have been progressively improved until they can retain water pressures of up to about 10 bars, and so can work at depths of almost 100 metres below water level. So when the French tunnels had to be driven through mixed wet strata near the coast, it was possible to specify TBMs to be

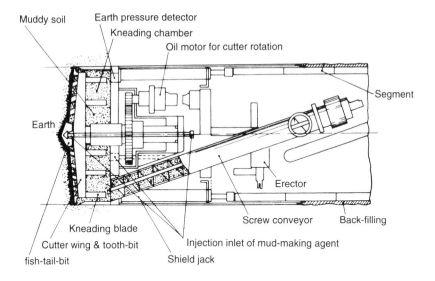

Muddy soil   Earth pressure detector
              Kneading chamber
              Oil motor for cutter rotation

Earth

Segment

Earth

Erector

Kneading blade
Cutter wing & tooth-bit   Screw conveyor   Back-filling
fish-tail-bit   Injection inlet of mud-making agent
                                     Shield jack

Fig. 5 Typical earth pressure balance tunnel boring machine.

capable of working in the 'closed' mode until the geological conditions improved sufficiently to make that unnecessary.

These machines were developed in the 1970s and 1980s. Meanwhile, in the USA, the emphasis has been on the development of hard-rock tunnelling machines. The Robbins Company in particular had developed more and more powerful TBMs to excavate self-supporting rock which did not require the immediate erection of tunnel linings.

The London digger shields, and all the other types I have previously described, propel themselves forward by hydraulic jacks which push off the segmental tunnel lining once erected close behind the TBM. The Robbins TBMs had no such lining to push forward off, so instead this company developed gripper pads to expand sideways and lock against the sides of the excavated rock tunnel. The softer the rock, the greater the area of gripper pad required to avoid damaging the rock.

The English undersea TBMs had an additional refinement. There was a telescopic section of the TBM which allowed excavation to be carried out independently of the erection of the tunnel lining segments. Thus, the forward section of the TBM, including the cutter head, was jacked forward off the gripper section, which was firmly anchored by the gripper pads. Meanwhile, the erection of a ring of tunnel segments proceeded independently behind the tail of the TBM. This system can only be made possible by the generally self-supporting nature of the chalk on the English side of the Channel, and even so, there had to be constant probing of the ground ahead of the TBM to ensure that no unexpected deterioration of conditions was encountered, and various provisions made, such as doors which could be closed to prevent inrush, emergency pumping arrangements, etc.

Tunnelling is often split into two categories: first, soft-ground tunnelling through clay, silt, sand, and gravel (to work through the particle sizes of 'soft ground') and secondly, rock tunnelling. The Channel Tunnel might be described as soft-rock tunnelling with the French forced to use soft-ground techniques and the British, fortunately, able to use elements of rock tunnelling techniques. I use the word British advisedly because although the tunnels were driven from England, many of the engineers were, as always, from Scotland, and there were large numbers also from all parts of the British Isles; but we should not forget that the accents of France, Austria, New England, and many other countries were to be found amongst those designing and building this huge international project.

There is not time to describe the back-up logistic arrangements created to construct these tunnels. Suffice it to say that the problems of ventilation, dust suppression, temperature control, fire-fighting, and power supply all over 20 km from the access point, were successfully overcome, and that the removal at the peak rate of several thousand tons of excavated material per hour, and the supply of men and materials to the tunnel faces were achieved by an underground railway system, albeit temporary, which exceeded in scale and complexity those of most of the capital cities of the world!

The tunnel linings

The instrumentation installed and left around the tunnels built in 1974–75 helped to give an unusually full understanding of the engineering properties of the chalk marl, and corresponding confidence in how to design the tunnel linings.

With such a uniform stratum of chalk marl it was possible to build up a mathematical model of the behaviour of the ground, with a combination of plasto-elastic and visco-elastic elements, and, despite the ability of the ground to stand unsupported for quite extended periods (depending on the size of the excavation), it was shown that it would be necessary to design tunnel linings sufficiently strong to withstand the total overburden pressure of the ground (and the sea) above the tunnel. On the English side, however, the low permeability of the ground would make it unnecessary to make the tunnel linings completely watertight, and it would therefore not be necessary for the linings to withstand the hydraulic pressure which could otherwise build up in the ground surrounding the tunnel. We tunnellers spend much of our lives trying to make our tunnels watertight, but here, paradoxically, we had to be equally careful not to allow the water to be kept out against its will in the coming decades. Instead, we had to be sure that what little water is finding its way in is deflected from doing damage to the plant in the tunnels, and is led away to the drains and the pumping stations.

So, the object was to design the most economical linings which could be built rapidly behind the advancing TBMs. A joint Anglo-French tunnel lining study, begun in 1986, soon concluded that precast concrete segments of high quality would suffice in all but the most arduous conditions. On the English side, the segments could, in the relatively dry conditions, be articulated and expanded directly against the cylinder of exposed chalk marl as it emerged behind the TBM, and all subsequent operations such as grouting could be carried out, hardly 'at leisure', but at least without delaying the advance of the TBM (Fig. 6).

The ring of segments 1.5 metres wide is built with mechanical erectors and then tightened against the ground by jacking in the closing 'key' segment. Pads on the backs of the segments bear directly against the ground; the remaining voids around the pads are filled with cement grout a few hours later.

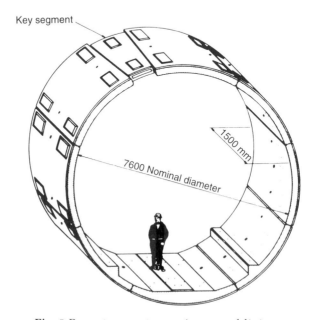

Fig. 6 Precast concrete running tunnel lining.

In the event, as has been mentioned, blocky and wet ground was met for some 3 km off the English coast, and the erection of this type of lining was not as straightforward as had been hoped, but it was made to work and for the whole of the rest of the drives was extremely successful.

In theory, the wider the ring and the smaller the number of segments in it, the more quickly the tunnel can be built. But segments become progressively more difficult to cast, and more vulnerable to damage in handling as they become bigger. The fewer the segments in a ring, the less flexible the ring and the greater the bending moment in individual segments when the load comes on. In retrospect it is fortunate that rings 1.5 metres wide were provided for the English tunnels as a wider ring would have entailed a larger area of exposed ground, and more overbreak near the coast.

The production of these precast concrete segments at a special factory on the Isle of Grain, North Kent, was an enormous project in itself. Large bulk carriers brought Glensanda granite from Scotland and about a million tonnes of cement has gone into the production of 450 000 segments, individually weighing between three-quarters of a tonne and eight tonnes. Tolerances had to be strict, especially on the bearing surfaces between segments, where enormous compressive loads have to be transmitted in exactly the right place, so the size of these great concrete units had to be accurate within one-tenth of a millimetre at the critical points. Very special arrangements are needed to measure with such accuracy, especially when other parts of the same unit do not need to be to the same high standard, and indeed it would have been very extravagant to have specified unnecessary accuracy.

As production neared completion, it was found that 1.6 per cent of segments had required some form of minor repair, and only 0.7 per cent had been rejected, including all early experimental work. Once production had settled down, the rejection rate was about one segment in five hundred.

The segments were taken to the work site by main line train and to the tunnel face by the temporary railways in the tunnels. Plate 14 shows a completed running tunnel.

The traditional tunnel linings for the most arduous conditions have been rings of bolted cast-iron segments. Until the 1930s, all of London's 'tubes' were built with grey cast-iron segments, as were the Blackwall and Rotherhithe tunnels under the Thames. The segments are as good as new after a century in the ground. Unlike mild steel, cast iron does not suffer seriously from corrosion, but it lacks the tensile strength of steel. In recent years a high grade spheroidal graphitic (SG) iron has been developed which is ideal for casting and also has the tensile properties of steel. This material is now used for tunnel linings which may suffer eccentric loads and bending stresses. One example of its use is to form openings in tunnels, which has always been a source of difficulty. In the past, the practice was to place temporary supports in the tunnel before building an opening. In the Channel Tunnel, the financial penalty of blocking the tunnel with temporary supports and paying a few days' interest on several billion pounds was so severe that a new method was devised, using intricate SG iron segments (Fig. 7) which could be built by the TBM as it swept past the future opening, but were strong enough to allow the hole to be formed by dismantling temporary segments.

For the TBM drives, a small reserve of SG iron rings was also kept in case there was an unexpected deterioration in ground conditions which would have made concrete segments inadequate.

In France, the tunnellers had to address the problem of producing a watertight tunnel from the start, and they settled on a strong, bolted, reinforced concrete grouted segmental lining with gaskets to make it watertight. They experienced teething troubles (as indeed we did in England), but they got through the difficult ground, and by that time their systems were working so well that they carried on with the same segments until they broke through to join the tunnels from England.

The method of junctioning was unusual. The British service tunnel drive, having run its course, was driven deliberately off to one side as

Fig. 7 Spheroidal graphitic iron segments forming an opening.

sharply as possible. The French service tunnel TBM was then driven close up to the British drive and a horizontal borehole was put through between the two, followed by a heading. There was therefore a transition length which allowed the small survey differences to be ironed out. The service tunnel was then completed by hand excavation, the French TBM was dismantled and removed, and the British TBM was simply concreted in, as to dismantle it would have caused expensive delay without benefit.

Similarly, with the later running tunnels, the British machines were 'left in', but it was decided that it would be even less trouble if they were driven steeply downwards and subsequently entombed in concrete poured down on to them. By that time, of course, all survey differences had been harmonized so the railway alignment would be perfect.

So much for the service tunnel and the running tunnels. There were also many other tunnels for sumps, pumps, electrical equipment, etc., and most impressive of all, the crossover tunnels. No railway company likes to operate without an ability to transfer trains to an adjacent track, either for routine maintenance or to bypass some difficulty or a broken-down train. So they have points and crossings so that any particular section of track can be isolated from the system. The Channel Tunnel is no exception, and the railway operations require two 'crossovers' in the undersea portion of the system, one about 7 km from the English coast, and the other off the French coast.

The crossover tunnels

Three supplementary boreholes, and detailed interpretations of a closely spaced geophysical survey, indicated an area near the desired position of the crossover where there was a comparative absence of even minor faulting, and a thickness of marl above the Gault clay which was adequate to accommodate the crossover structures. That allowed the brave decision to be taken (after most careful reviews of the design and of the risks) to build a single large cavern to provide the crossover (which could otherwise have been built in a complicated series of junction chambers).

The following operations had to be carried out, all fitting in with the overall construction schedule and without delaying the critically important progress of the service tunnel and running tunnel TBMs (Fig. 8):

1. The service tunnel had to be driven towards the future site of the crossover cavern. A path was chosen through the rather less desirable glauconite marl close above the Gault clay. Once well below the future running tunnels, the tunnelling swung towards the North

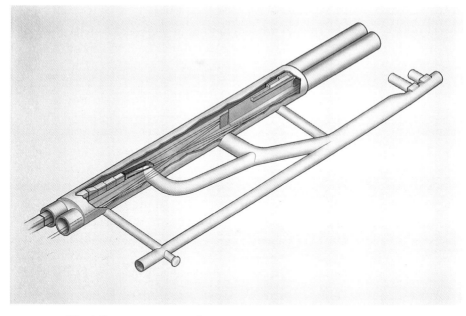

Fig. 8 Crossover cavern showing access tunnels from the service tunnel in the foreground.

and the service tunnel was driven past the future crossover, parallel to it but below it. Once past the crossover, the reverse procedures were carried out to get back on to the usual station between the future running tunnels and continue towards France.

2. Construction adits had to be excavated from the service tunnel to the site of the crossover, and an undersea cavern was constructed of record size, 21.2 metres wide, 15.4 metres high, and over 150 metres long: this is as wide as a cricket pitch is long and as high as three double-decker buses stacked on top of each other. It was excavated in a series of drifts, each of a safe size; the chalk marl being explored as excavation proceeded and being strengthened by spraying concrete on the exposed excavated surfaces, by fixing rock bolts and, most importantly, by monitoring movements of the ground, and additional strengthening was provided accordingly. All this had to be achieved without delaying the service tunnel TBMs progress towards France.

3. The ends of the cavern had to be strengthened and it had to be made ready to receive the running tunnel

TBMs, which converged to enter it as close together as could be risked (leaving a pillar of ground only 2 metres thick between them).

4. The TBMs had to be pulled through the cavern, taking the opportunity to refettle them, and then relaunched almost 8 km on their way towards France.

5. The permanent waterproof lining of the cavern had to be completed without interfering with the through traffic servicing the TBMs still on their way to meet the French.

6. The permanent railway equipment had to be installed.

The interaction between the design, the construction, and the scheduling of the work was unusually intricate, and the success of the whole episode is a great achievement.

To stand in the vast fully excavated cavern, with work continuing, 7 km offshore with the sea-bed about 35 metres above one's head, was an awe-inspiring experience.

The French crossover is similar in final operational concept, but their programme dictated that the two running tunnels should be driven through the site of the crossover before it was constructed, and the less favourable geological conditions led to their forming the cavern by driving a series of headings, each filled with reinforced concrete on completion, around the perimeter of the cavern, thus forming a very strong structural ring around the cavern without at any stage having a large unsupported tunnel (Fig. 9). They could then safely excavate the remainder.

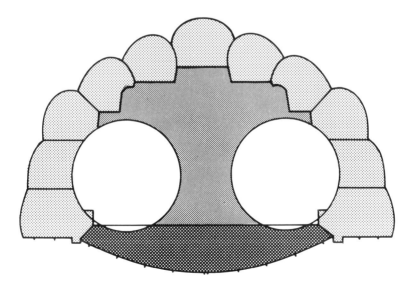

Fig. 9 A cross-section through a French crossover section.

Conclusions

When I wrote the synopsis for this article I felt all fired up about the way this project was organized. I am a 'public works' man, a phrase which embraces a great tradition of providing such things as tunnels, railways, highways, ports, and power stations to governments or occasionally to private corporations, with the minimum of fuss and the maximum of competition. Why, then, this extraordinary method of giving one consortium a kind of copyright to the railway tunnel solution which had been developed over the previous decades, bullying another consortium to prepare a vast highway bridge/tunnel scheme, encouraging a third consortium to put forward a strange tunnel which was to carry both railways and motor vehicles, presumably at different times, in which a puncture might lead to death by carbon monoxide poisoning! And then to choose the Eurotunnel solution, the one we all know to be the obvious solution: in effect, the one cancelled by a previous government only 10 years earlier.

And then, in order to avoid committing a penny of taxpayers' money on such a risky project, to involve scores of banks, hundreds of lawyers and consultants, and thousands of shareholders, all but the last highly paid, to take the risks—when many of the risks are political and created by the governments, not the operators!

If the tunnel operators do a good job, the profitability of the project is largely in the hands of those who control national railways, those who govern passport procedures, those who decide on duty-free allowances, those who have to deal with terrorism, or French farmers, those who influence currency exchange rates, and those who control rates of interest.

But in thinking about it, I mellowed. After all, Mrs Thatcher and Monsieur Mitterrand actually got the show on the road and kept it there—something which all others had failed to do for over a century. International projects are inevitably dreadfully complicated: all the more credit to those who, despite being enmeshed in immense organization charts, are still managing to complete this great project.

Perhaps the involvement of private finance was the only way of appeasing the Treasury.

Since Monsieur de Gaumond dived into the Channel and Colonel Beaumont of the Royal Engineers developed his TBM, many engineers have given the best years of their lives to the Channel Tunnel, and I am indeed fortunate to have seen the tunnels successfully completed from England to France. Many of us involved in the project have only spent a very small part of our time in the tunnels, including myself. So let me take this opportunity of paying a tribute to those who worked down there for many long shifts.

JOHN V. BARTLETT, CBE, MA, F. Eng., FICE

Born in 1927 he was educated at Stowe and Trinity College, Cambridge, with honours degrees in Mechanical Sciences and in Law. Has been engaged since 1951 on heavy civil engineering work first with John Mowlem & Co. Ltd., contractors, and then with Mott, Hay, & Anderson, consulting engineers, for whom he was senior partner and chairman 1975–1989. He has been Chairman of the British Tunnelling Society (1977–9), President of the Institution of Civil Engineers (1983–4), and his papers and inventions have brought him two Telford gold medals and other prizes from the Royal Society, the Institution of Civil Engineers, and the American Society of Civil Engineers. He and his firm (now Mott MacDonald) have been involved in the various schemes for the Channel Tunnel for many years, and designed the English tunnels now being completed.

Bioremediation: helping nature's microbial scavengers

R. R. CHIANELLI

On Good Friday, 27 March 1964 a massive earthquake of greater than eight on the Richter scale occurred in the Chugash mountains just North of Alaska's Prince William Sound. Spreading from its epicentre the quake rocked Prince William Sound and its coastline in a 500-mile arc from far out in the Aleutian Islands to Juneau just North of Vancouver, British Columbia.[1]

In that same year a consortium of oil companies led by British Petroleum and Esso Petroleum, worried over their vulnerability demonstrated in the recent Suez Crisis, started serious exploration on Alaska's North Slope leading to the discovery of a giant oil-field that would eventually supply 25 per cent of the liquid energy production of the United States. The discovery of this oil led to a national debate about how the oil might be brought to market. In 1974 Vice President Spiro T. Agnew cast a vote to break a tie in the US Senate passing legislation that allowed a pipeline to be built from the North Slope through 800 miles of the Alaskan wilderness terminating in Valdez, North America's most northern year-round ice-free port.[2]

The 1964 earthquake was the most powerful in North America since the turn of the century, eclipsing even the great San Francisco earthquake of 1910. In Anchorage the earth dropped 30 feet in the middle of the city. In Prince William Sound at places such as Knight Island, coastlines were lifted as much as 3 feet in a few moments, creating the vertical rock walls seen today. But the real devastation occurred shortly after the quake as huge tidal waves crashed into the coast. One such wave surged up the Valdez arm of Prince William sound, building in size as it traversed the 30 miles from the mouth into the cul-de-sac containing the small port of Valdez. Reaching the end, it smashed into the town, spending itself against the ring of mountains encircling the harbour. It ruptured fuel oil

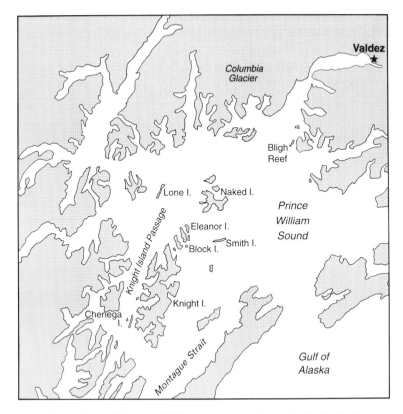

Fig. 1 Map of Prince William Sound and the Gulf of Alaska.

storage tanks, filling the harbour with oil, and swept over 70 people to their death. Ironically, 25 years later, as the people of Valdez were preparing to commemorate the disaster, the *Exxon Valdez* left the terminal at Alyeska shortly after loading 1.3 million barrels of Alaska North Slope crude oil and travelling the 30 miles to the mouth of the Arm, went off course, and ran hard aground on the appropriately named, Bligh Reef. It was early in the morning, Good Friday, 24 March 1989.

The tanker had eight of its 13 tanks holed but, in a remarkable feat of seamanship, was stabilized allowing almost 80 per cent of the cargo to be removed by smaller tankers such as the *Exxon Baton Rouge*. This was indeed fortunate as during this process a fierce storm struck which severely endangered the tanker, threatening to make the spill approximately the size of that from the *Amoco Cadiz* that had broken up in a similar storm off the coast of Brittany in March of 1978 in what was the largest tanker spill ever.[3]

The storm also hindered further efforts to contain the spill and the spreading oil slick moved slowly out into Prince William Sound past

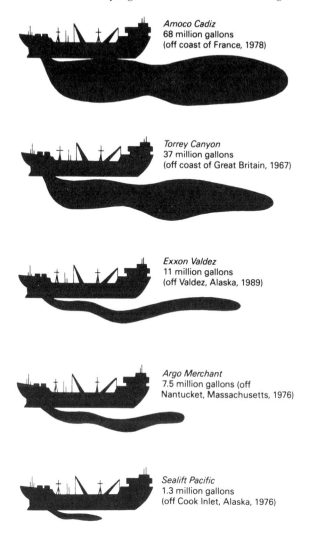

Amoco Cadiz
68 million gallons
(off coast of France, 1978)

Torrey Canyon
37 million gallons
(off coast of Great Britain, 1967)

Exxon Valdez
11 million gallons
(off Valdez, Alaska, 1989)

Argo Merchant
7.5 million gallons (off
Nantucket, Massachusetts, 1976)

Sealift Pacific
1.3 million gallons
(off Cook Inlet, Alaska, 1976)

Fig. 2 Magnitude of oil spills across the world. Reproduced from
the USEPA's *Research summary: oil spills* (1979).

Naked Island and Knight Island and down into the Gulf of Alaska, impact-
ing about 1300 miles in total—or about 15 per cent—of the area's
shoreline.[4] The now empty, stricken tanker was towed to Naked Island
where temporary repairs were made and from there she was towed to San
Diego where permanent repairs were made.

The clean-up: summer 1989

As the slick moved from Bligh Reef toward the beaches, heroic efforts
were made by workers, local fishermen, and Exxon to protect sensitive

areas such as salmon hatcheries. All such areas in the path of the slick were protected. In Prince William Sound itself about 486 miles of shoreline were impacted, or less than 20 per cent of the total shoreline, a difficult number to estimate exactly given the convoluted nature of the shoreline.

It was in this context that I attended a meeting on 5 April 1989, called within Exxon Research and Engineering Company, to find novel clean-up technology based on sound scientific principles. As a volunteer I, like many Exxon scientists and engineers, felt badly about the spill and wanted to do something about it. I was, nevertheless, quite surprised to find myself in Alaska one month later heading Exxon's bioremediation research task force. Earlier we had decided that nutrient-enhanced bioremediation might be a valuable clean-up tool and a short time later the United States Environmental Protection Agency (USEPA) came to the same conclusion. In May of 1989 I travelled to Valdez to participate in discussions that would lead to a joint USEPA/Exxon program to field-test bioremediation techniques for potential application in the shoreline clean-up.

I will always remember my first view of the oiled beaches as we surveyed them from the air. All the impacted areas were extremely remote with no roads, so access was only by boat, plane, or helicopter. At the height of the clean-up effort over 11 000 workers, 1100 vessels, and 84 aircraft were in service. To me it looked as I imagined the invasion of Normandy might have looked (the invasion of Normandy had about 1400 vessels, but much larger ones of course).

The first step in cleaning the beaches was to wash them with sea water. Initially cold water was used in a deluge system that easily removed the major portion of the freshly landed oil. As the summer progressed the oil became more viscous and warm water was used in pressurized spray systems. Out of approximately 486 miles in Prince William Sound cleaned in this manner, most were vertical rock outcrops and 74 miles were sloped beaches.

You can imagine how easy it was to clean oily rocks with just warm water! Although most of the oil was physically removed by this process, the rocks were still covered with a thin layer of oil. Something else was needed. The answer was lurking in the sea just below the pine trees which overhang most of the vertical rock beaches in Prince William Sound. For millions of years pine trees have been dripping hydrocarbons called terpenes into the sea. It is the terpenes which give the pine its distinctive smell. Microbes called oxidative hydrocarbon degraders have evolved to consume these molecules. Terpenes are similar to the heavier molecules found in petroleum. This 'molecular soup' has made the bacterial consor-

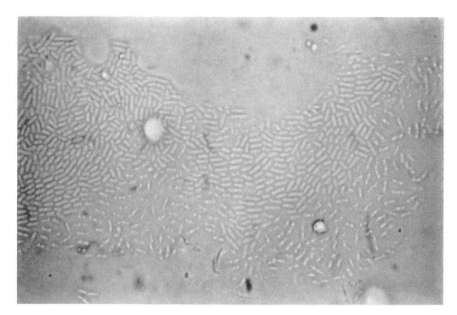

Fig. 3 Optical micrograph of Prince William Sound hydrocarbon degraders. Each organism shown is approximately two microns in length.

tium of Prince William Sound particularly potent in its ability to degrade various hydrocarbon molecules. The functioning of consortia of this type that contain hundreds of different types of microbes, performing different functions and yet acting collectively, is in itself, a fascinating problem facing modern microbial ecologists. Some of these microbes can easily be grown in the laboratory as seen in Fig. 3.

Because of the intrinsic activity of the indigenous microbes, addition of micro-organisms was not necessary and probably would have given no additional benefit because the organisms introduced would not have adapted to the Prince William Sound environment and thus would not have survived. The safety of adding 'foreign' organisms was also another unknown. Addition of organisms was one strategy that was initially proposed and later proven to be ineffective although the technique may have future importance in specific situations.

Before the spill the hydrocarbon degraders accounted for approximately 0.1 per cent of the total microbial population. This number was limited by the amount of hydrocarbon normally present which is the food source for the bacteria. After the spill this percentage jumped to 40 per cent because the hydrocarbon limitation was removed. Hydrocarbon degraders also require oxygen to convert the hydrocarbon molecules to CO_2 and H_2O.

This process, called oxidative biodegradation, can be thought of as a 'slow combustion' of the hydrocarbon. On the porous rocky beaches of Prince William Sound oxygen is always abundant, being constantly replenished by the tidal ebb and flow of oxygen-rich sea water.

The micro-organisms now become limited by another factor. For approximately every carbon atom of petroleum converted to CO_2, another carbon atom is used to build more biomass. More hydrocarbon-degrading organisms means more consumption of hydrocarbons and yet more hydrocarbon degraders and so on. Thus, the process is self-accelerating and is termed autocatalytic. However, the microbes require nitrogen and phosphorus to build more of themselves and the low natural concentration of nitrogen and phosphorus limited more rapid biodegradation after the spill. The biomass that results from this process becomes food for higher organisms that are themselves consumed by yet higher organisms. Therefore the petroleum is safely converted atom by atom to make more of the species that exist naturally in the food chains.

Nutrient-enhanced bioremediation

Given that the hydrocarbon degraders were nutrient limited it was in principle a simple matter to add nutrients to the beaches in much the same manner as you might fertilize your garden. This fertilization process, called 'nutrient-enhanced bioremediation', had been previously known but never tried on such a large scale.[5] There was, however, an important difference to consider: the nutrients must be kept out of the water column to the greatest extent possible to avoid algal blooms as algal growth would also be enhanced by the nutrients. Nutrient run-off from farmlands has been implicated in large algal blooms such as those that have recently occurred in the Adriatic Sea.

We thought that run-off reduction might be accomplished by using special nutriation strategies. One such strategy was to use commercially available 'slow release' agricultural fertilizers, which are designed to release nitrogen and phosphorus over periods of weeks, thereby keeping the concentration of the nutrients relatively low at all times. A second strategy was to deliver to the beach with a sprinkler system just the right amount of nutrient in soluble form at each low tide. The proper amount of nutrient delivered in this manner was the estimated amount that the micro-organisms could use in one cycle. Finally, a third strategy was considered, the use of an oleophilic or 'oil-loving' nutrient.

The only oleophilic fertilizer commercially available at the time of the spill was Inipol EAP-22. This product was developed by researchers at Elf-Aquitaine after the *Amoco Cadiz* spill in 1978.[6] Inipol is a pale yellow liquid micro-emulsion, containing both nitrogen and phosphorus, that can be applied by spraying. Initial laboratory studies had shown that the nutrient limitations that existed in Prince William Sound could be overcome by utilizing the above strategies and that by doing so hydrocarbon degradation would be significantly accelerated.

However, formidable barriers had to be overcome if we were to use these techniques in the summer of 1989. We had started in April and we needed to develop scientific underpinning, all of the application strategies and put the logistics structure in place by the start of the summer. We knew that weather would shut down operations in the middle of September, giving us a very narrow window of opportunity for application of nutrients. To address these issues, parallel scientific, engineering, and application programmes were put in place. Furthermore, we needed to get approval from the various state and federal agencies which required that we prove in the field that nutrient-enhanced bioremediation was safe, effective, and feasible. It was this last task that proved to be the most formidable barrier of all because many in Alaska were highly sceptical of any proposals put forth by Exxon.

In order to expedite the development of bioremediation techniques we formed a partnership with the USEPA. In May this partnership was formalized with the institution of a USEPA/Exxon field programme to test the above concepts.[7] On 1 June 1989, the field programme was initiated at Snug Harbour on Knight Island in Prince William Sound where oleophilic and slow release fertilizers were applied to test plots. Control plots were also included. Later at Passage Cove on Knight Island the sprinkler system was evaluated. The USEPA also began extensive toxicity tests in the waters directly off the test beaches and in the laboratory.

By 4 July 1989, a remarkable result was apparent on the oleophilic test plot. A white window appeared within the borders of the treated area after 2 weeks, while the control beaches remained unaffected. Initially, only the surface of the beach was cleaned but as time went on the process continued down from the surface. After approximately 8 weeks the sprinkler and slow release beaches began to show similar, though less dramatic results. Furthermore, toxicity tests and monitoring results offshore showed that there were no adverse effects from the treatments. On the basis of these results conditional approval was obtained for application to begin.

Full-scale application began on 1 August 1989. Over 70 miles of beaches were treated with Inipol in Prince William Sound. These were all of the beaches that were suitable (the remainder of the shoreline was vertical rock, which was cleaned with hot water). Heavily oiled beaches were first washed with water, while lightly to moderately oiled beaches

R. R. Chianelli

Fig. 4 The 'magic window' in Snug Harbor on Knight Island in Prince William Sound. The result of oleophilic fertilizer (Inipol-EAP22) application after 4 weeks.

were treated directly. The Inipol was transported on small barges with storage tanks and sprayed on the beach by an operator carrying a hand-held sprayer. Within several weeks dramatic results were seen on the treated beaches that were similar to those observed on the test plots. Surface cleaning was initially observed followed by deeper cleaning as time proceeded.

By the middle of September, beaches all across Prince William Sound were showing similar effects and the project became the world's largest and most successful bioremediation project. Bioremediation techniques were also extended to oiled beaches in the Gulf of Alaska. In cases where subsurface oil only was present a commercially available slow release fertilizer called Customblen (Sierra Chemicals Inc.) was used.

The winter 1989–90 laboratory programme

The dramatic effect on the cleanliness of the beaches which nutrient-enhanced bioremediation had shown, led to an intensive laboratory programme in the winter of 1989–90 to generate a better understanding of the results and to answer some questions that would arise in considering additional bioremediation treatments for the spring of 1990. Some of

Fig. 5 Spraying of oleophilic fertilizer on a beach in Prince William Sound.

the questions regarding the safety and efficacy of nutrient enhanced bioremediation were:

1. What was the mechanism of the action of Inipol in clearing the beaches?
2. What were the products of nutrient-enhanced accelerated bioremediation?
3. Was nutrient-enhanced bioremediation effective on subsurface oil?
4. How much could biodegradation be enhanced with nutrients?

Further questions were also asked about the toxicity of Inipol to local species of marine life. This question was answered by the USEPA in a separate programme not discussed here. They found that when Inipol was properly applied at correct rates, toxicity was not a problem. These studies formed the basis for application guidelines first used in the summer of 1990.

Returning to the questions regarding the safety and efficacy of nutrient-enhanced bioremediation, it had been easy to show in the preliminary experiments mentioned above, using simple shake flasks, that Inipol accelerates bioremediation. In these experiments oil, sea water, and nutrients were mixed in Ehrlenmeyer flasks in appropriate proportions.

These shake flask experiments were designed to look for the hydrocarbon degrading activity of Prince William Sound organisms in a rich medium (Bushnell–Haas broth, familiar to biologists and rich in nitrogen, phosphate, and potassium, supplemented with 3 per cent NaCl, Wolfe's trace elements, and Pfennig's vitamins) containing everything required for growth on 1 per cent weathered crude oil. Artificially weathered Alaskan North Slope (ANS) crude oil was used. The 'artificial weathering' consisted of removing, via distillation at 521 °C, 30 per cent of the lightest hydrocarbons to simulate the composition of the oil that actually landed on the beach. Two different inocula were used: apparently oil-free sea water collected from an oiled beach in Prince William Sound, and effluent from the Alyeska water treatment facility that treats tanker ballast water in Valdez. Three millilitres of inoculum were added to 25 ml of medium, and 280 μl of oil were added on top. After 10 days of shaking at 15 °C, oil was no longer floating on the surface of the medium. The contents of each flask were extracted three times with 5 ml of dichloromethane, and the extract was dried with anhydrous Na_2SO_4. Gas chromatographic (GC) analyses indicated that up to 60 per cent of the oil was degraded in the Alyeska treatment and the sea water inoculum during this period of time. There was no measurable degradation in un-inoculated control experiments. HPLC (high pressure liquid chromatography) analyses of the residual oil showed that both inocula reduced the saturates (paraffins and saturated cyclic hydrocarbons) and increased the aromatic percentage of the residual oil relative to the starting oil composition. Gas chromatography (GC) analyses of the residual oil revealed a dramatic disappearance of the paraffins from the oil, and sulphur-specific GC detection demonstrated significant reduction or removal of some sulphur-containing species. From these experiments it was concluded that indigenous bacteria from Prince William Sound are capable of degrading many of the components of weathered Prudhoe Bay crude oil. These experiments also indicated the effectiveness of soluble nutrients in enhancing biodegradation. Similar experiments concluded that:

1. Inipol accelerates biodegradation of ANS crude oil as determined by GC analysis.

2. Acceleration of biodegradation increases with increasing Inipol concentration from 0 to 40 per cent.

3. Inipol-accelerated bioremediation degrades more than 60 per cent of the oil under laboratory conditions.

4. Inipol-accelerated bioremediation increases with increasing temperature and decreases with decreasing temperature between 2 and 20 °C.

5. Inipol-accelerated bioremediation cleaned Prince William Sound oiled rocks only through biodegradation and not through physical means.

Some of the conclusions listed above can clearly be seen in Figs 6 and 7. Figure 6 shows the cleaning effect of Inipol in the presence of hydrocarbon-degrading organisms and Fig. 7 shows clean rocks which resulted from the biological action of Inipol accelerated biodegradation. The results of the flask studies are in agreement with earlier French studies which reported 70–80 per cent biodegradation of Arab light crude using Inipol.[8]

Three other types of experiment of increasing complexity were undertaken to extend the results of the previously described shake flask experiments and to approach conditions found on the beaches:

1. Full fate flask experiments: these larger scale experiments were designed to determine the carbon mass balance of oil and degraded oil products including biomass and CO_2 in a closed system.

2. Small beach microcosms: these experiments were designed to simulate tidal cycling with nutrient replenishment.

3. Large beach microcosms: these 36-inch high, 12-inch diameter microcosms with full tidal simulation and analysis were designed to confirm results of smaller microcosms and address subsurface bioremediation.

Fig. 6 Comparison of treated and untreated rocks from Ingot Island in Prince William Sound. (a) Rocks treated with oleophilic fertilizer. (b) Untreated rocks.

Fig. 7 Rocks on Knight Island approximately 3 weeks after treatment with Inipol-EAP22

The full fate experiments were designed to capture and measure all products that were produced by accelerated biodegradation. Three litre flasks loaded with oil, sea water, and nutrients were constantly stirred and kept at 15 °C. Air was sparged into the flasks at 100 cm³/min and the CO_2 evolved during biodegradation was collected and weighed on a daily basis. At various points in the experiment samples were collected and analysed to determine the chemical composition of the unbiodegraded oil, the composition of any dissolved molecular products, and the amount of biomass produced. The reference oil was weathered ANS crude oil as described above. The products of biodegradation as measured in these experiments were approximately 50 per cent biomass and 50 per cent CO_2 (water is also a product but was not determined). Intermediate products were also detected but only at p.p.m. levels because they are degraded more quickly than the starting oil. The results are in agreement with many literature studies and the earlier cited French work on Inipol EAP22. Further, nutrient-enhanced biodegradation produces exactly the same products as natural biodegradation. Chemical analysis of the remaining oil yielded information about the chemical composition of the oil during this process. The changing molecular composition during biodegradation

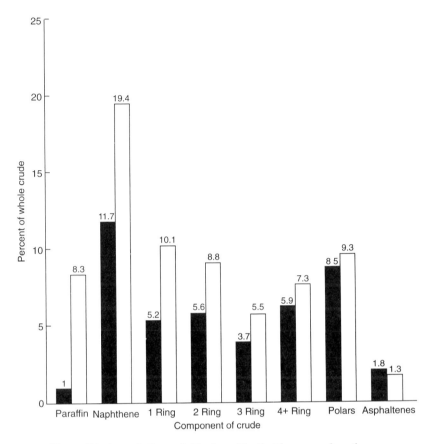

Fig. 8 Biodegradation of Alaskan North Slope crude oil components. Solid bars, after biodegradation; open bars, ANS 521F.

is shown in Figure 8, it can be seen that all of the molecular components are biodegraded with the exception of the asphaltenes, which represent less than 2 per cent of the crude oil. Figure 8 also shows that the larger, more complex molecules are biodegraded more slowly than the smaller molecules. This is in good agreement with the work of Oudot who measured the biodegradation rates of the components of Arabian light crude.[9]

Even though probably underestimating the rates in open systems, total biodegradation rates can be estimated from CO_2 evolution data of the type shown in Fig. 9. After approximately 16 days it was found that the CO_2 evolution begins to slow down, presumably due to product inhibition in the closed system. Initial uninhibited CO_2 evolution rates, calculated from time zero to 16 days during which period the evolution of CO_2 versus time is roughly linear, are plotted in Fig. 10 against the applied nitrogen concentration. The plot yields a straight line which passes

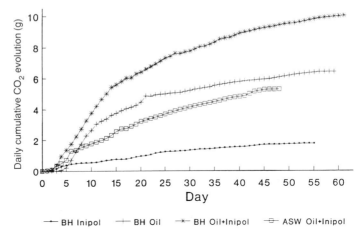

Fig. 9 Evolution of CO_2 with time in full fate flask experiments. BH, Bushnell–Haas medium; ASW, artificial sea water.

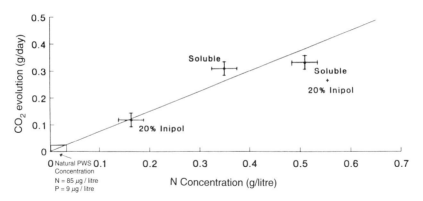

Fig. 10 Evolution of CO_2 against nitrogen concentration from data in Fig. 9.

through the origin, from which one can estimate a 16 day rate at the natural nutrient concentration. This calculation, though far too simple (ignoring washout of nutrient for example) shows that the natural rate would require 500 days to deliver the same amount of nutrient through tidal cycling as a 10 per cent Inipol application; i.e. the natural rate is slower by a factor of 30. Of course these estimates await experimental confirmation in the field.

In an effort to introduce the elements of tidal cycling and nutrient replenishment into the experiments, a series of small microcosm experiments was performed. Using modified Soxhlet extractors as microcosms, samples of Prince William Sound collected in October 1989 with an oil loading of 5 g of oil/kg rocks were placed in an extractor.[10] Artificial sea water containing appropriate levels of nitrogen, phosphorus, and trace nutrients was then cycled in an accelerated manner (1 cycle per hour). The sea water was 50 or 80 per cent changed on each cycle depending on the series (two series) of the experiments. The entire apparatus was kept at 15 °C in a cold room. Complete chemical analyses are in agreement with the full fate flask experiments described above, indicating no essential change in the basic chemistry in going to an open system with oiled rocks. In addition when a bacterial inhibitor was added (100 mg/litre $HgCl_2$) no conversion was seen and no oil was removed in repeated tidal cycling. After 22 days significant amounts of oil actually fell off the Inipol-treated rocks during removal, suggesting an additional mechanism for removal of oil during accelerated biodegradation. We know from other studies that the clay sediments from Prince William Sound affect oil removal from the rock through the process of flocculation.[11] This process is enhanced by polar components in the oil that chemically are strongly bound to clay surfaces. As biodegradation proceeds the content of polar components increases which in turn accelerates the natural beach cleaning process. The undegraded oil leaves the beach as a flocculated packet containing microbes, partially degraded oil, nutrients, and fine particles. It continues to be biodegraded as it moves into the water, probably rapidly because of its high surface area, clay-suspended state. This process is exactly the same as the natural process but is accelerated by high nutrient concentrations.

The microcosms received fresh sea water every day, with levels of nitrogen and phosphorus equivalent to those found in Prince William Sound sea water. These low levels of nutrients are not enough to sustain biodegradation in the closed system of the full fate flasks, but with constant replenishment they are sufficient to sustain the natural biodegradation in Prince William Sound and in the control microcosms. The amount of biodegradation seen in the microcosms thus provides another estimate of the stimulation of the natural biodegradation rate by the addition of Inipol. The control microcosms lost 0.1 g of a 1.9 g initial load in 30 days, while the samples with Inipol lost 0.65 g of starting material. This 6.5-fold stimulation is probably an underestimate of the potential stimulation in the field, since the nutrient-treated microcosms were almost certainly partially oxygen-limited due to their design.

In order to investigate biodegradation on a larger scale and to investigate subsurface biodegradation, large-scale microcosms were constructed.[12] These microcosms were rock-filled columns that have a diameter of 1 foot and a depth of 3 feet. Oxygen uptake during biodegradation was measured as a function of time and depth. Custoblen alone,

Inipol alone, Inipol plus Customblen, and an aqueous fertilizer solution were investigated. The cumulative oxygen uptake of a 10 per cent Inipol treatment was measured and judged by oxygen uptake for approximately 30 days. Addition of Customblen re-established the oxygen uptake. The data indicate that a rate acceleration of a factor of 5 was achieved in this experiment. In addition these experiments indicated good biodegradation at depth. The results support EPA beach observations which indicate that Inipol bioremediated to a depth of at least a foot by visual observation and that in soluble spray applications nutrients penetrated to at least 7 feet in porous beaches.[13]

The beaches of Prince William Sound and the Gulf of Alaska were monitored in parallel with the laboratory studies. Monitoring consisted of biological and chemical analyses of samples from over 20 beaches in Prince William Sound and the Gulf of Alaska. Both the biological monitoring and chemical analysis showed that bioremediation had been enhanced by the 1989 fertilizer application. This was in spite of the fact that there was only one application of fertilizer, which occurred at the end of the summer of 1989. The median number of oil-degrading bacteria on the bioremediated beaches from September to January was found to be approaching 100 times the number found on untreated beaches. The increase on bioremediated beaches persisted until January even though the total number drops during the winter months. The 49 samples measured for the September bioremediated samples can be divided into surface and subsurface samples. Enhanced numbers are also found in the subsurface on bioremediated beaches. During this same period of time, on the same beaches, the number of heterotrophs remained constant.[14]

Prince William Sound hydrocarbon degraders are extremely active in the ambient conditions of the Alaskan seas.[15] Using aerobic enrichment culture techniques, several hundred microbial isolates from Prince William Sound have been collected since the oil spill. All isolates identified to date are members of bacterial or fungal genera already known to encompass hydrocarbon-utilizing species. The isolates were predominately aerobic, Gram-negative, psychrotrophic organisms with the possible presence of some microaerophiles. Species isolated have included: *Acinetobacter calcoaceticus*, *Alcaligenes* sp., *Arthrobacter/Brevibacterium* sp., *Flavobacter/Cytophaga* sp., *Oceanospirillum* sp., *Pseudomonas fluorescens*, *Pseudomonas putida*, *Pseudomonas stutzeri*, *Pseudomonas* sp., *Pseudomonas vesicularis*, *Trichosporon* sp., and others. The majority of the isolates have yet to be identified or could not be identified based on the biochemical and morphological tests used. However, the data obtained place the isolates into the following groups: *Acinetobacter/Moraxella*, *Alcaligenes/Achromobacter*, *Arthrobacter/Brevibacterium*,

Flavobacter/Flexibacter/Cytophaga, Pseudomonas, Spirillum, and others. With few exceptions, the isolates demonstrated hydrocarbon-degrading ability in both freshwater and marine environments. All isolates grew from 5 °C to room temperature. Only a few isolates grew well at 2 °C. All of the isolates grew on either tetradecane or hexadecane. A few isolates tested on aromatic substrates demonstrated growth on phenan-threne or 1-methylnaphthalene. Experiments utilizing radiolabelled model compounds indicate that the hydrocarbon-degrading community degrades oleic acid, hexadecane, and phenanthrene simultaneously in the presence of crude oil. Oleic acid was consumed more rapidly than the other model compounds.

In Newport Beach, California on 1–2 February 1990, a shoreline clean-up workshop was attended by representatives of Exxon, EPA, NOAA (National Oceanographic and Atmospheric Administration), USCG (United States Coast Guard), and ADEC (Alaskan Department of Environ-mental Conservation) to discuss the results of summer application field tests and laboratory tests. This group advised the USCG on clean-up technology that they then supervised. On the basis of data from field tests, laboratory tests and actual application, the workshop attendees agreed to the following conclusions regarding nutrient-enhanced bioremediation as applied in Prince William Sound:

1. Nutrients increase rate of biodegradation of oil.

2. Visual effects and laboratory studies indicate fertilized plots show accelerated biodegradation over control plots. Oil chemistry from field tests also supports this conclusion.

3. Oil residue removed is in a highly degraded state.

4. When properly applied, Inipol's mechanism is bio-degradation and not chemical rock cleaning.

5. No adverse ecological effects were detected from the fertilizer applications.

6. Fertilizer application is best used to take advantage of the short time window when the water temperatures are warm enough for optimal microbial activity.

7. Accelerated biodegradation through fertilizer applica-tion is an important option in the consideration of methods to further reduce oil residues in Alaska.

8. Fertilizer application shows good prospects for bio-degradation at depth, since soluble nutrients penetrate to depth and Inipol was effective to a depth of 1 foot.

9. Both gravimetric and chemical field data were highly
    variable.

Bioremediation in 1990

Armed with the above conclusions I went back to Alaska in March of 1990. The purpose of the trip was to join with the USEPA in a tour to visit many of the villages and towns on the perimeter of Prince William Sound to convince the population of fishermen and native Americans that bioremediation was safe and effective. I found the trip to be not only exciting with volcanic eruptions and earth tremors, but also extremely informative. It taught me first-hand the dilemma faced by any scientist interested in applying new technology. This dilemma pits the modern public's fascination with new technology against their fear of anything that they do not understand. Nutrient-enhanced bioremediation was greeted initially with great scepticism. There was a mistrust of anything that involved adding chemicals to the environment. Fortunately, this fear was usually accompanied by a genuine curiosity about scientific questions surrounding issues concerning the environment of Prince William Sound. It was this curiosity coupled with hard work that eventually overcame the initial scepticism of most people. The lesson for the scientist is quite clear: be open and honest regarding your scientific findings and take the time to explain them clearly to anyone with interest. Also, try to convey your own love of science and avoid being 'political'. Most importantly we had a good story. Nutrient-enhanced bioremediation was safe and effective and would significantly accelerate the recovery of Prince William Sound. Following this proscription we were able to overcome the scepticism and approval was granted in April of 1990 for additional application of nutrient-enhanced bioremediation techniques.

Once bioremediation was approved as a treatment method of choice, the appropriate fertilizer could be selected. The condition of the beach was the major contributing factor to the selection of fertilizer. When a beach exhibited a surface cover, coat, or stain in sufficient concentration to merit application, Inipol was applied at the rate of 0.06 lb/ft^2 (17 g/m^2). When no surface oiling was apparent and yet there was evidence of subsurface oil, Customblen alone was applied at the rate of 0.03 lb/ft^2 (93 g/m^2). These rates are based on the results of EPA toxicological and eutrophication analyses.[7] Whenever fertilizers were applied, passive visual devices were put into place on the beaches for a minimum of 24 hours following completion of the work to deter wildlife from entering the treated area while minimizing wildlife stress. These devices include scare-eye balloons, Mylar tape, scarecrows of Mylar sheeting, animal

Table 1

	1989 application	1990 first application	1990 second application
Inipol			
gallons	71 300	17 100	10 200
sq. ft	9 765 000	2 300 000	1 400 000
Customblen			
lb	19 400	80 800	38 900
sq. ft	3 680 000	4 700 000	3 400 000

silhouettes, and roped flagging. A comparison of the 1989 and 1990 fertilizer application is indicated in Table 1. It can be seen from Table 1 that much more Customblen was applied in 1990. This reflected the fact that beach surfaces were much freer of oil in 1990 and the major object of treatment was the subsurface oil which surveys indicated was present in many beaches. A visual monitoring programme over the summer indicated that 84 per cent of the beaches had improved with the beaches receiving both Customblen and Inipol showing the most improvement.

A bioremediation monitoring programme was developed to follow results from the spring 1990 re-application. This effort included scientists from US EPA, ADEC, and Exxon. The studies focused on quantifying three effects of treatment in order to demonstrate the safety and effectiveness of bioremediation:

1. The degree of biodegradation was assessed by both microbiological and chemical criteria. Bacterial counts were determined using standard procedures, and oil samples were collected from the beaches to measure the amount of biodegradation.

2. The toxicity associated with the application of fertilizers was measured by collecting water samples when the tide had covered the application area and by using that water in toxicity tests on mysid shrimp.

3. Nutrient loading in the water from treated beaches addressed the potential for stimulating algal growth. The amount of nutrients was measured in the water, and a chlorophyll measurement was used to examine algal presence.

One test beach in this program was on Knight Island, where a combination of Inipol/Customblen application was approved for mid-May application. The preliminary results of monitoring this beach were as follows.[16]

1. The activity of oil-degrading bacteria in surface sediments and subsurface sediments sampled at a depth of 30 inches has been enhanced 2- to 3-fold and sustained for 32 days post-application.[17]

2. By employing ratios of degradable and undegradable fractions of the oil components, an estimate of the baseline oil degradation rate was obtained. On a heavily oiled low energy shoreline that did not receive bioremediation in 1989, the rate has been approximately 2.3 g/year in the subsurface.

3. Fertilizer application has not resulted in any adverse ecological effects.

After the summer of 1990 an extensive investigation commenced of the changes in oil chemistry that occurred on treated and untreated beaches. This effort was hindered by the large degree of variability that was observed in the chemical analysis of beach samples. However, by using internal 'biomarkers' such as hopane molecules[18] and by exhaustive statistical analysis it was conclusively shown that nutrient addition enhanced biodegradation by a factor of 3–15.[19] The implication of this rate is that the recovery of Prince William Sound was significantly accelerated by the treatments. Experts have predicted that an untreated oil spill can take 10–20 years for natural recovery.[20]

Analysis of the chemistry data went further, though. It has been shown that all beaches, whether treated or untreated, followed a model that showed that the rate of biodegradation is primarily related to the amount of nitrogen delivered per unit of oil present. This means that in the future, nutrient-enhanced bioremediation techniques may be even more effective as the treatment frequencies and amounts are optimized based on the experience in Alaska.

Conclusion

Experience in Alaska has shown that nutrient-enhanced bioremediation is an effective new tool for the clean-up of spilled oil. In 1989 over 70 miles of beaches in Prince William Sound were treated with nitrogen-containing nutrients. This was followed by further treatment in the summer of 1990. Bioremediation of Alaskan beaches was effective for both surface and subsurface oil in conjunction with water washing to remove bulk oil. In addition it was demonstrated that nutriation of beaches presented no unacceptable environmental risk.

Laboratory experiments and simulations proved to be an important link in establishing the effectiveness of bioremediation given the difficulty and expense of demonstrating bioremediation effects in the field. Research in the future will include efforts to strengthen further the basis for using laboratory data to determine the appropriateness of bioremediation techniques in specific situations. Though the need for nitrogen in the Alaska case is clearly established, further work is required to establish the fundamental relation between the observed statistically significant nitrogen dependent models observed for the treatment of beaches in Alaska and known theoretical microbiological models. An increase in understanding the underlying principles of nutrient-enhanced bioremediation should lead to further improvements in the application of these techniques in mitigating the effects of oil spills.

Acknowledgements

The author wishes to acknowledge the contributions of the following individuals for their efforts in the monitoring and/or research programmes: T. Aczel, R. E. Bare, M. Bhalla, D. Elmendorf, D. Ferrughelli, G. N. George, M. W. Genowitz, M. J. Grossman, C. E. Haith, M. J. Hicks, T. Jermansen, F. J. Kaiser, L. G. Keim, S. R. Kelemen, R. J. Kennedy, P. Kocian, A. Kurs, P. J. Kwiatek, A. I. Laskin, R. R. Lessard, S. R. Lindstrom, R. Liotta, J. R. Lute, R. L. Mastracchio, S. J. McMillen, V. Minak-Bernero, M. A. Modrick, F. Pfeifer, W. K. Robbins, A. F. Ruppert, G. W. Schriver, J. Senius. Special thanks to R. M. Atlas, J. R. Bragg, S. M. Hinton, J. Phillips, R. C. Prince, E. I. Stiefel, and J. B. Wilkinson for useful and stimulating conversations.

References

1. Graves, W. P. E. (1964). *National Geographic*, **126**, 112.
2. Yergin, D. (1991). *The Prize*. Simon and Schuster, New York.
3. Gundlach, E. R., Boehm, P. D., Marchand, M., Atlas, R. M., Ward, D. M., and Wolfe, D. A. (1983). *Science*, **221**, 122.
4. Owens, E. M. (1991). *Shoreline conditions following the Exxon Valdez Spill as of fall 1990*. 14th Annual Arctic and Marine Oilspill Program. Woodward Clyde, Seattle.
5. Atlas, R. M. and Bartha, R. (1973). *Env. Sci. Technol.*, **7**, 538.
6. Ladousse, A. and Tramier, B. (1991). *Proceedings of the 1991 Oil Spill Conference* (San Diego). API, p. 577.
7. Pritchard, P. H. *et al.* (1991). *Alaska oil spill bioremediation project science advisory board draft report*, USEPA/600/9-91/046a.
8. Sirvins, A. and Angles, M. (1984). *Biodegradation of petroleum hydrocarbons*, NATO ASI series, Vol. G9.
9. Oudot, J. (1984). *Marine Env. Res.*, **13**, 277.
10. Chianelli, R. R. *et al.* (1991). Bioremediation technology development and application to the Alaskan spill. *Proceedings of the 1991 International Oil Spill Conference*, American Petroleum Institute, Washington DC, pp. 549–58.

11. Jahns, H. O., Bragg, J. R., Dash, L. C., and Owens, E. H. (1991). *Proceedings of the 1991 Oil Spill Conference* (San Diego). API, p. 167.
12. Bragg, J. R., Prince, R. C., Wilkinson, J. B., and Atlas, R. M. (1992). *Bioremediation for shoreline cleanup following the 1989 Alaskan oil spill.* Exxon Company, USA.
13. Pritchard, P. H. and Costa, C. F. (1991). *Env. Sci. Technol.*, **25**, 372.
14. Prince, R. C. *et al. Appl. Microbiol.* submitted.
15. Zobell, C. (1973). In *Microbial degradation of oil pollutants* (ed. D. G. Ahearn and S. P. Meyers). Publication no. *LSU-SG-73-01*, Center for Wetland Resources, Louisiana State University, Baton Rouge, p. 153.
16. Prince, R. C., Clark, J. R., and Lindstrom, J. E. (1990). Report submitted to the USCG.
17. Lindstrom, J. E. *et al.* (1991). Microbial populations and hydrocarbon biodegradation potentials in fertilized shoreline sediments affected by the T/V Exxon Valdez oil spill. *Appl. Env. Microbiol.*, **57(9)**, 2514–22.
18. Elmendorf, D. L., Hinton, S. M., Chianelli, R. R., and Prince, R. C. (1991). Oil spill bioremediation. In *Proceedings of the Conference on Hazardous Waste Research* (ed. by L. E. Ericson). Kansas State University, Manhattan, KS.
19. Bragg, J. R., Prince, R. C., Harner, E. J., and Atlas, R. M. (1993). Bioremediation effectiveness following the Exxon Valdez spill. In *Proceedings of the 1993 International Oil Spill Conference*, American Petroleum Institute, Washington DC, pp. 435–47.
20. Baker, J. M. (1992). *CONCAWE*, 117.

RUSSELL R. CHIANELLI, Ph.D

Born 1944 in Newark, New Jersey, he joined Exxon in 1974 after receiving his Ph.D. in chemistry from the Polytechnic Institute of Brooklyn. He is currently a Section Head of the Catalytic and Biological section and Environmental Sciences Coordinator of Exxon's Corporate Research Laboratory in Clinton, New Jersey. He was also the Research Task Force Leader for Exxon's Alaskan Bioremediation Project. He is Past President of the Materials Research Society of the United States and is author of over one hundred publications and patents. His interests are history, wine, and automobiles.

Our genes under the microscope

H. J. EVANS

Advances in our scientific knowledge depend upon the generation of new ideas, or of new insights, and the development and the application of new technologies, not to mention a great deal of patience, sweat, and sometimes tears. In this article I hope to avoid the sweat and tears, but I may ask for a little patience, particularly from those who will be familiar with some of the work that I am going to describe.

With the enormous developments in human genetics over the last decade it has become increasingly clear that much of human abnormality and ill-health is, in large part, a consequence of the abnormalities, or mutations, in the genes and chromosomes that we inherit from our parents. Moreover, such mutations are also continuously arising in the somatic cells of our bodies and some of these mutations are causal factors in the genesis of cancers. The genes I am going to talk about therefore are your genes and my genes and I should begin by addressing the question, 'what are genes and chromosomes?'

Genes and chromosomes

Our genes are the *bits* of information that we inherit from our parents, and pass on to our children, which specify and make up the blueprint for the development of a normal or indeed abnormal human being. The number of different genes in a single human genome (i.e. the genetic information in each cell) is probably of the order of 50 000 and is unlikely to be more than 100 000 and each nucleated cell in the body has exactly the same full complement of genes. Each human sperm or egg therefore contains at least 50 000 different bits of information. Each person has duplicate copies of these ~50 000 genes, one set from each parent. The paternally derived and maternally derived copies of a given gene are not always identical, though. For example, a gene specifying eye colour, say brown eyes, may be the same from each parent or it may be slightly different, one parent passing on a gene specifying brown eyes but the same gene from the other parent may specify blue eyes.

Each gene comprises a segment of deoxyribonucleic acid, DNA, and is located at a fixed and specific site on a specific chromosome, which is a thread-like structure a few micrometres in length that consists of one double-stranded molecule of DNA that is tightly packed and runs from one end to the other. Each of us has 46 chromosomes or 46 duplexes in all the cells of our bodies except the germ cells. In the formation of a sperm or egg cell the number is halved so that each gamete contains 23 chromosomes (Fig. 1). The 46 duplex molecules of DNA in a human nucleus, if aligned end-to-end, would give a thread some 5 feet in length and around 15 Å in diameter.

The DNA is essentially a twisted ladder with rungs, each made up of two complementary DNA bases, linked to a phosphate–sugar backbone; the DNA in a single human sperm nucleus, i.e. a haploid genome, contains around 3×10^9 base pairs. Each gene is simply a segment of DNA with a specific sequence of base pairs that code for a specific protein. There are but four different bases and a specific sequence of three bases provides the coding unit (codon) for a single amino acid: the nature of the

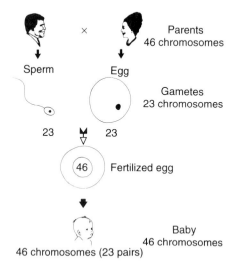

Fig. 1 The chromosome cycle. Each sperm contains 23 chromosomes and around 50 000 genes. Each egg has also 23 'partner' chromosomes and the same genes, but for any one of these 50 000 genes there may be small but very significant differences between the paternal and maternal forms.

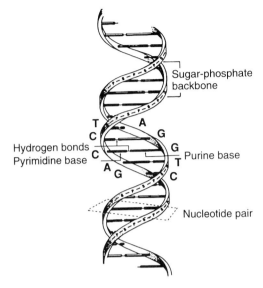

Fig. 2 Portion of a DNA molecule showing the four kinds of nucleic acid bases, known by the abbreviations A, T, C, and G, and their paired arrangements within the DNA duplex. T residues always pair with A residues, and C with G. A is therefore complementary to T and C to G. Thus when the two strands of the double helix separate, each can be used as a template to make a copy of the original double helix.

codons and the sequence of their arrangement along the DNA strand specifies the protein product of a gene (Fig. 2). There are some 20 different amino acids and each of our different proteins may consist of one or more of a series of linked amino acids (peptides). A protein may have many thousands of amino acids and it is the nature and order of these units that specify its structure and function.

At one extreme, the gene coding for a small protein, or peptide, such as a growth hormone or a neuropeptide, may consist of around 1×10^3 base pairs; at the other extreme, the gene coding for the protein dystrophin, which in an abnormal or mutated form results in Duchenne muscular dystrophy—a severe and eventually lethal disease—consists of almost 2×10^6 base pairs. Informational DNA of the sort that makes up our genes probably amounts to no more than 10–20 per cent of the DNA in a human nucleus, so that the bulk of the DNA in our genomes is non-coding DNA. Most of this non-coding DNA consists of short lengths of repeated sequences which cannot code for proteins but which are present in hundreds or even millions of copies in the genome—DNA that is sometimes irreverently referred to as 'junk DNA'. Our genes can therefore be pictured as coding islands in a sea of non-coding DNA, each island having a fixed position at a specific site on a specific chromosome.

Individual genes are too small to be visualized under a light microscope, but the chromosomes on which the genes are located are readily visible. Human chromosomes can be obtained from small lymphocytes which comprise a major fraction of the white cells in peripheral blood. Normal blood contains something like 1×10^6 small lymphocytes per millilitre; these cells normally do not divide, but if exposed to an antigen they will enter into a mitotic cell division, which makes their chromosomes easier to see.

Standard microscope slide preparations giving well spread and flattened chromosomes stained with a dye, readily enable us to count up the normal 46 chromosomes in a human nucleus under the light microscope and allow us to pair up the chromosomes; one member of each pair being a copy of that chromosome derived from the father's sperm and the other being a copy of that chromosome derived from the mother's egg. With this technique we cannot distinguish all of the chromosomes individually, but can only characterize them on the basis of overall shape and size. However, we can readily see that in the normal female each member of each pair is identical to its partner, but that in the male the two chromosomes in one of the 23 'pairs', the sex chromosomes, are not identical, there being one 'X' chromosome and a much smaller, male-determining, 'Y' chromosome. Females have two X chromosomes. The normal male and female chromosome constitutions are thus 46,XY and 46,XX respectively. The father can contribute either an X or a Y chromosome to a single sperm, while the mother will always contribute an X chromosome to each of her eggs.

The chromosomes at the time of cell division (mitosis) are double structures each consisting of two strands, chromatids, as a consequence of replication of the DNA in interphase, the 'resting' stage between all divisions. Each chromosome contains a constriction called the centromere which holds the chromatids together until they separate at the time of cell division so that when the cell divides into two, half of each chromosome passes into a daughter cell so that each daughter cell receives 46 chromatids which then replicate and so on.

We can look at these chromosomes at a higher magnification, for example using the scanning electron microscope where each chromatid appears to have a surface resembling looped carpet pile and its length appears to be segmented. If we digest away some of the proteins associated with the DNA of the chromosome and examine it under the electron microscope then we can see that the DNA is present as loops attached to a protein scaffold. The chromosome, however, is not a uniform structure

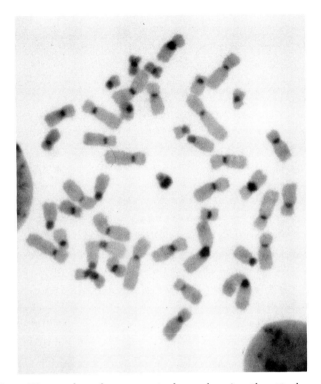

Fig. 3 Human lymphocyte metaphase showing the 46 chromosomes following staining with a technique to highlight the large
blocks of repetitive non-coding DNA. The cell is from a normal
male and the chromosome in the centre is the Y chromosome
which consists largely, but not entirely, of non-coding DNA.

for by using various techniques we can reveal that different types of DNA
are located in different parts of the chromosome.

If we expose the chromosomes to a very alkaline salt solution and stain
with Giemsa stain we reveal large staining blocks (C-bands) predominantly
at the centromere and over most of the whole of the long arm of the Y
chromosome, which we can show consists of non-coding repeated DNA
sequences (Fig. 3). The amount of this C-band DNA differs between
people. For example, some men have a very small Y chromosome with
very little of this kind of DNA, while others have very large Y chromosomes with the extra length being due entirely to additional repeated
DNA. This non-coding DNA on the Y chromosome is of no genetic consequence: men with small Y chromosomes are just as virile and fertile as
men with large Y chromosomes.

The use of a slightly different technique shows us that each chromosome is made up of a series of bands of DNA, the sizes and positions of

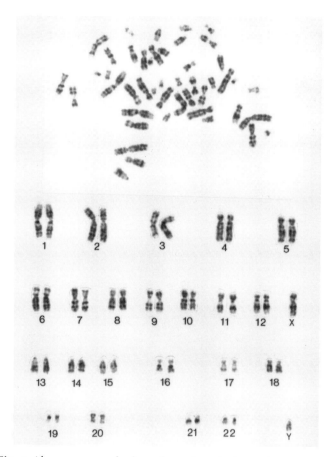

Fig. 4 Above: a spread of condensed metaphase chromosomes from a lymphocyte cell stained to show the specific chromosome bands. Below: paste-up of the chromosomes above (a 'karyotype') arranged to show that they represent 23 pairs of chromosomes which are identical in overall morphology and banding pattern, except for the X and Y chromosomes. This spread can thus be identified as coming from a male. One of each member of a pair is an exact copy of that chromosome that was present in the mother's egg or the father's sperm.

which are specific for each specific chromosome so that we can readily identify each of the 23 pairs of chromosome in the complement (Fig. 4). Around 1000 bands are visible in the long chromosomes in the early prophase of cell division (Fig. 5), but as the chromosomes condense during division we can resolve around 400–500 bands.

How do these bands relate to our genes?

There are a variety of approaches to detect the presence and pin-point

Fig. 5 Relatively uncondensed chromosomes, seen at an earlier stage of mitosis relative to those in Fig. 4 and showing the large number of transverse bands that are characteristic of each particular chromosome.

the locations of genes that are beyond the scope of this article. But if the DNA base sequence is known, or part of a sequence that specifies a gene, or the sequence of a segment of the amino acids that form part of the protein coded by a gene, then copies of that segment of the gene can easily be made by linking together the correct bases in the appropriate order in the laboratory using a machine called an oligonucleotide synthesizer. In this way we can synthesize complementary probes to a segment of a gene. If we label these probes with a radioactive label, or better still a fluorescent label, then these probes can, under specific conditions, bind to that particular complementary DNA sequence and only that sequence in a chromosome preparation and then we can see where that sequence or gene is located. Using these techniques, and other approaches, we can see that most of our genes are located in the light staining bands of the chromosome, the heavily staining bands again consisting largely of non-coding DNA.

The above-mentioned technique of detecting specific DNA sequences by the use of fluorescent tags on DNA probes is referred to as FISH, an appropriate acronym which stands for fluorescent *in situ* hybridization. The basic FISH technique involves hybridizing chromosome preparations, in which the chromosomes have been denatured by acid to make their DNA single-stranded, with complementary DNA probes. The probes are labelled with biotin, the presence of which can be detected by a biotin-specific antibody binding agent—avidin—which is itself detected by an antibody labelled with a fluorescent molecule such as fluorescein isothio-cyanate. The end result, for example, is that if we use a DNA probe that is complementary to the terminal ends, or telomeres, of chromosomes, the ends will fluoresce. We have centromere-specific probes that will reveal all the centromeres and others that are specific for certain centromeres. Moreover, we can actually use chromosome-specific probes to 'paint' individual chromosomes and of course there are also probes specific for single genes. For example, a probe complementary to part of the gene responsible for kidney cancers in children picks out the Wilms' tumour gene on the short arm of chromosome 11. We now have probes for a large number of human genes, which has made it possible to map the position of these genes in the genome.

I have discussed above some of the fundamentals concerning our genes and chromosomes, but of particular interest is the fact that quite a large proportion of ill-health in humans is a consequence of the inheritance, via our germ cells, of mutations (i.e. abnormal chromosomes or abnormal genes), or a result of gene or chromosomal alterations occurring spon-taneously or induced by environmental agents in our somatic cells. Such changes in somatic cells are responsible for the development of cancers, but first let us look at mutations in germ cells.

Germ cell mutations

Mutations that we inherit from our parents are broadly of two sorts, those that affect whole chromosomes or parts of chromosomes and which we can readily see under the microscope, and those that affect single genes and which we can only visualize if we have a specific DNA probe. The distinction between the two is sometimes arbitrary. For example, the con-sequences of a mutation affecting a single gene can result either from a single base change in one codon in the coding sequence of that gene, or from the disruption of that gene by a breakage of the DNA within the gene; such breakages are often visible under the microscope.

Chromosome mutations that we can detect under the microscope turn out to be not uncommon. All chromosomally normal individuals have 46 chromosomes, but there are a number of well-known inherited abnormali-ties that are a consequence of the inheritance of an abnormal number of chromosomes. The best known example is perhaps Down's syndrome, which is a result of the inheritance of an extra copy of chromosome number 21 and has a prevalence of around one in every 800 live births. Other chromosomally abnormal conditions which are equally frequent among males are an additional X chromosome, giving the 47,XXY constitution of Klinefelter's syndrome—a condition resulting in tall, often slender, sterile males with some female secondary sex characteristics, and an additional Y chromosome, resulting in 47,XYY males who are larger than the average male, but are fertile and have, on the average, a slightly lower IQ. Additional X chromosomes are also found in some females, e.g. 47,XXX, but with a lower frequency; they are often associated with mental retardation.

In addition to these chromosome gains, or indeed losses, a small proportion of individuals may inherit a chromosome that is broken, or has a part deleted or moved elsewhere (translocated). A good example is the deletion of part of the short arm of chromosome 11 which predisposes the child who inherits this to the development of Wilms' tumour of the kidney—often associated with the lack of an iris in the eye and mental retardation (Fig. 6).

How frequent are these conditions?

If we examine chromosomes—under the light microscope—from placental blood from a series of consecutive live births, as we have done in Edinburgh, it turns out that one in every 160 live newborn babies is chromosomally abnormal. This frequency of chromosome abnormality is very much higher (~5%) if we look at still births and higher still (~50%) in fetuses which abort. It is evident from these studies that over one-half of early abortions are brought about because of the presence of a chromosome anomaly. It is only those chromosome mutations that are compatible with viability that come through to term and affect the eventual liveborn individual. Evidently chromosome mutations in human gametes are of frequent occurrence (Table 1).

In recent years we have been able to develop techniques that allow us to look at chromosomes within human sperm. The technique involves presenting the sperm with hamster eggs stripped of their outer covering; the sperm then fertilize these eggs and display their chromosomes. A number of laboratories have been using this technique to look at the chromosomes in human sperm and the results show that between 2 and 12 per cent of sperm from a normal fertile healthy male are chromosomally abnormal—a very high frequency of chromosome mutation. Although we have these techniques for revealing the chromosomes in human sperm it is far more difficult to analyse directly the chromosomes in human eggs. However, the development of techniques to make test-tube babies has enabled us, indeed required us, to look at the chromosomes of human eggs fertilized *in vitro*. Studies on cells, blastomeres, of very early human embryos in

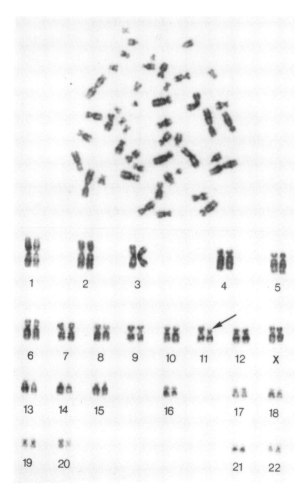

Fig. 6 Metaphase spread (above) and karyotype (below) of lymphocytes of a girl who had inherited an abnormal chromosome 11, which lacked part of the short arm (arrowed below), from one of her parents. The girl developed Wilms' tumour of the kidney as a consequence of this constitutional genetic deficiency.

Edinburgh show that chromosome abnormalities, that is new mutations, are present in at least one of every 10 fertilized eggs. The human species therefore has a remarkably high rate of chromosomal mutations in its germ cells and it transpires that the frequency of chromosome anomalies that are passed on from parent to offspring is very much dependent on the ages of the parents at the time of conception.

The chromosome constitution of a fetus can be ascertained by analysing the fetal cells that are normally shed into the surrounding amniotic fluid

Table 1. The frequencies of major chromosome abnormalities in spontaneous abortions, perinatal deaths, and live births (all per 1000)

Anomaly	Live births (n = 43 558)	Perinatal deaths (n = 500)	Spontaneous abortions		
			<28 weeks (n = 941)	<18 weeks (n = 255)	<12 weeks (n = 1498)
Polyploidy	0.13	2.00	53.13	101.96	160.21
Autosomal trisomy	1.24	28.00	151.98	250.98	330.44
Monosomy X	0.046	2.00	72.26	156.86	93.46
Other	4.43	24.00	27.62	39.22	30.70
Totals	6.26	56.00	304.99	549.02	614.82

Note that over 60 per cent of early abortions are associated with a chromosomal mutation and that constitutional chromosome abnormalities are found in one in every 160 live newborn babies and in 5 per cent of still-births. Polyploidy, having three or more complete sets of chromosomes, as opposed to the normal two, that is 69 or more chromosomes rather than 46; autosomal trisomy, having three copies of a non-sex (autosomal) chromosome as opposed to the normal two copies; monosomy, having a single copy of a chromosome, in this case an X (sex) chromosome.

and which can be readily sampled from the pregnant female at around 16 weeks' gestation. The results of such an analysis on the European population show how strongly dependent is the prevalence of chromosome abnormality on maternal age. Similarly we can also demonstrate an age dependence effect on the paternal gametes for a variety of single gene mutations that result in abnormalities or disease. The reasons for this increased frequency of mutations in gametes with increasing parental age are not fully understood, but we cannot exclude the possibility of a contribution from exposure to environmental mutagens such as radiations, alcohol, combustion products, etc.

Somatic cell mutations

Exactly the same kinds of mutational events that we see in chromosomes and genes in germ cells also occur in somatic cells. If a somatic mutation occurs in a proliferating cell in a renewing tissue then the mutation is passed on to its descendent cells which will all then possess the mutation. These mutations may involve the loss or gain of whole chromosomes, deletions of large parts of chromosomes, large rearrangements within and between chromosomes, or small deletions, rearrangements, or base substitutions within a gene. These mutational events occur quite frequently. For example, if we examine the peripheral blood lymphocytes of a normal healthy human male or female, then some 1–2 per cent of the cells may have a missing, or contain an additional, chromosome and a similar number may contain chromosomes with deleted segments or structurally rearranged chromosomes. Like mutations in germ cells, the frequencies of these somatic mutational events increase as we grow older.

These mutational events that we observe occur spontaneously, but their frequency is very much increased if we expose ourselves to mutagens or carcinogens such as ionizing radiations or ethylene oxide, vinyl chloride, or certain pesticides. We can readily demonstrate large increases in chromosomal mutations in blood cells of patients receiving high doses of X-rays as part of a cancer therapy programme, but we can also detect significant increases in people exposed to radiations, at levels within the accepted international limits, as part of their occupation.

Some years ago a Belgian colleague, wishing to confirm and extend our Edinburgh studies showing increased chromosomal damage in occupationally exposed radiation workers, looked at chromosome aberration frequencies in blood cells of workers in a nuclear power plant and compared these with two kinds of controls. The first were office workers and the second workers in a fossil fuel power plant. The results confirmed an increase in chromosomal mutations in blood cells of the workers in the nuclear power plant, but showed an even greater increase in chromosomal mutations in the blood cells of the fossil fuel power plant workers. This is not surprising as it is well known that the combustion products of hydrocarbons are highly mutagenic and carcinogenic, so that we get a bigger mutational effect at low levels of exposure to reactive chemicals than we observe with ionizing radiation.

Most of the spontaneous and mutagen-induced mutations that we see in blood cells, bone marrow cells, or skin cells are of little or no consequence. For example, a mutation that involves, for instance, a gene coding for a pigment in the eye will be quite irrelevant if that mutation occurred in a liver or blood cell. Similarly, a mutation in a differentiated muscle cell— which is no longer dividing—may be of little consequence, but a mutation occurring in a proliferating cell at a gene locus involved in the control of proliferation, or differentiation, may have dire consequences—this is what happens in the initiation and development of human cancers.

In recent years it has become increasingly evident that chromosomal and gene mutations are the causal factors involved in the initiation and development of human neoplasms. The first clear evidence for this was the discovery by Peter Nowell and David Hungerford in Philadelphia some 25 years ago of a characteristic chromosome abnormality called the Philadelphia, or Ph[1], chromosome that was present in the malignant cells of patients with chronic myeloid leukaemia (Fig. 7). It turns out that the Ph[1] chromosome is a translocated chromosome number 22, which has

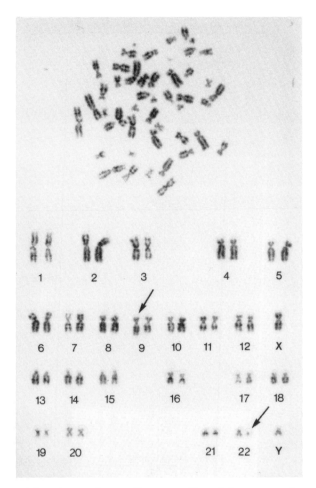

Fig. 7 Metaphase of a malignant cell and karyotype from a patient with chronic myeloid leukaemia showing the Ph[1] or Philadelphia chromosomal mutation. The mutation occurs only in the malignant cells and involves an exchange of parts between a chromosome 9 and a chromosome 22 (see Fig. 8).

exchanged its terminal region with a slightly smaller terminal region of chromosome 9. At the site of the exchange on chromosome 9 there is a gene, called the *abl* gene, whose protein product forms an early part of the cascade of events involved in driving cells into a proliferative state. The consequence of the translocation is to alter the protein product of the *abl* gene so that the mutated gene results in the cells being reprogrammed to undergo continued proliferation. The normal *abl* gene is an essential gene that is required for normal development, but in its altered—mutated— form it becomes a positively acting cancer gene or oncogene (Fig. 8).

t 9:22(q34;q11) in CML and ALL

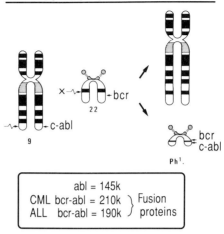

abl = 145k
CML bcr-abl = 210k ⎫ Fusion
ALL   bcr-abl = 190k ⎭ proteins

Fig. 8 Diagrammatic representation of the translocation resulting in the Ph[1] chromosome in chronic myeloid (CML) and acute lymphocytic (ALL), leukaemias. The resultant juxtaposition of the *abl* and *bcr* genes results in the formation of an altered *abl* gene and an altered 'fusion' protein; the points of chromosome breakage and exchange are slightly different in the two forms of leukaemia and result in different modifications of the *abl* protein.

There are now very many examples of specific chromosome rearrangements involving specific genes in specific cancers. As a second example, I would like to refer briefly to a translocation between chromosomes 14 and 18 which is found as a specific rearrangement in follicular B cell lymphomas. The function of B lymphocytes is to make immunoglobulins (antibodies), so the immunoglobulin gene on chromosome 14 is very active in these cells. The translocation results in placing a gene called the *bcl2* gene, which is normally on chromosome 18, closely adjacent to the immunoglobulin gene; the *bcl2* gene then becomes subject to the same control as the immunoglobulin gene, making it extremely active. The B cells are programmed to undergo an eventual cell death and be removed as normal turnover from the lymphatic system. However, the increased activity (up-regulation) of the *bcl2* gene prevents the normal programmed cell death, so these cells continue to proliferate and result in a tumour, a lymphoma.

The two examples to which I have referred identify but two of at least 50 genes in the human genome which are essential for normal development and body functions, but which are sometimes referred to as proto-oncogenes since following mutation, or up-regulation, they act as cellular oncogenes. Mutations in these genes then can be viewed as positive, or dominant, events abnormally driving or maintaining the cells through uncontrolled proliferation. There is, however, a second class of cancer genes, called tumour suppressor genes or anti-oncogenes, which have the reverse property. In their normal active state the function of these tumour suppressor genes is to act as controlling brakes on the cell cycle, preventing cells from proceeding into division. I will now turn to consider this second category of genes.

The first evidence for the presence of tumour suppressor genes came from studies on hybrid cells formed by the fusion of a cancer cell with a normal cell. If human cancer cells are injected into an immunodeprived mouse they will grow to give a human tumour in the mouse. If, however, a human tumour cell is fused with a normal cell and these hybrid cells are injected into a mouse, they will not give rise to a tumour. The tumorigenicity is suppressed by the introduction of a normal genome, i.e. normal chromosomes. If these hybrid cells are grown in culture then they will start shedding what are essentially surplus normal chromosomes from the donor. Eventually, at some point in time, the hybrid cells which have lost donor chromosomes regain their tumorigenicity. When the chromosomes of these hybrid cells that have reverted to their original malignant state are analysed, it is found that they have always lost one particular chromosome, or part of that chromosome; this shows that there is a gene on that chromosome which, when present, suppresses malignancy—a tumour suppressor gene.

Many of you will be aware of families in which cancer, or rather the predisposition to cancer, appears to run as a familial trait. It transpires that these familial cancers are almost always a consequence of the inheritance of a defective, or absent, copy of a tumour suppressor gene. The classic examples here are the childhood tumours of the retina (retinoblastoma) and the kidney (Wilms' tumour).

Retinoblastoma exists in two forms, familial and sporadic. In the familial form tumours develop in both eyes and each eye may have multiple tumours. In the sporadic form only one eye is affected, usually with only a single tumour. If we look at the peripheral blood cell chromosomes of a child with the familial form of the disease then we can show, either by observation under the light microscope or through the use of a DNA probe, that the child has inherited one normal chromosome and one defective chromosome 13 which lacks a segment in the region called band 13q14. The tumour no longer has a normal chromosome 13, but has two copies of chromosome 13, both defective at the q14 locus.

The explanation for these findings is as originally proposed by Al Knudson, namely that in the familial form, the child inherits one mutation of a retinoblastoma gene on chromosome 13, but that a second somatic mutation of the retinoblastoma gene on the other chromosome 13 is necessary for a tumour to develop. In the non-familial form of the disease, two independent somatic mutations within one cell are necessary; these are rare co-incident events and hence affect one eye and give rise to a single tumour.

The retinoblastoma gene has been isolated and cloned and we now know a great deal about its function. It acts as a normal and controlling brake preventing cells from entering into mitosis. We can readily show that it has a tumour suppressing function when present in its non-mutated form, for retinoblastoma tumour cells introduced into immuno-deprived mice give rise to tumours, but this tumorigenicity is abrogated if a normal retinoblastoma gene is introduced into these cells.

Retinoblastoma is but one of a number of inherited cancer predispositions that are the consequence of the inheritance of a mutated tumour suppressor gene followed by a later somatic mutation of the partner tumour suppressor gene. Many of the cancers that are clearly seen to be predisposed to by inheritance are relatively uncommon cancers and this is one reason that they stand out as being inherited cancers. What of the more common cancers of lung, colon, breast, and ovary? It turns out that for a proportion of all of these cancers, people may also be predisposed to them by the inheritance of a mutated tumour suppressor gene. Our own work in Edinburgh has shown that there can be predisposition for around 20 per cent of colon cancers, a very common cancer, at least 5–10 per cent or more of the common breast and lung cancers, and an even higher proportion of ovarian cancers. In all these cancers an initial initiating event is often a mutation in a tumour suppressor gene. This is then followed by the activation of one or more dominant oncogenes and by mutations in genes coding for components of the cell surface so that these malignant cells may escape and metastasize to other parts of the body. We now know that tumours evolve by a series of genetic changes that involve specific types of genes. The best understood of these changes are those involved in the emergence of colon cancer where we are at last beginning to understand the nature and sequence of genetic changes that are responsible for this disease.

In this article I began by introducing you to your genes and I have attempted to show how mutations in genes contribute in large measure to human ill-health. In some cases it is clear that we may be able to circumvent the development of catastrophic abnormalities due to the transmission of mutated genes, but I have not had time to discuss the new potential of somatic gene therapy for those who suffer from genetic disabilities. Potential there certainly is, and certain forms of gene therapy to treat inherited diseases, such as cystic fibrosis, are almost with us. I hope I have also been able to provide you with at least a flavour of some of the exciting developments in cancer research which have underlined the importance of genetic changes and which now offer us the prospect of new and powerful approaches not merely to detect early cancers, but to give us new, and hitherto unthought of, ways to develop effective therapies.

Further reading

Weatherall, D. J. (1991). *The new genetics and clinical practice*, 3rd edn. Oxford University Press.
Vogel, F. and Motulsky, A. G. (1986). *Human genetics*, 2nd edn. Springer-Verlag, New York.

H. JOHN EVANS, Ph.D., F.I.Biol., F.R.C.P.(E.), F.R.C.S.(E.), F.R.S.E

Born 1930 in Llanelli, South Wales, educated at Llanelli Boys Grammar School and the University College of Wales, Aberystwyth. He joined the MRC Radiobiology Research Unit at Harwell in 1955 and was head of the Cell Biology Section. In 1964 he moved to the Chair in Genetics at the University of Aberdeen and in 1969 moved to his current post as Director of the MRC Human Genetics Unit in Edinburgh where he also holds an honorary chair at the University. He has been involved in a wide variety of research projects in human genetics and cytogenetics, has a special interest in the effects of radiations and chemicals in inducing mutations and cancers, and has published over 250 papers and edited a number of books and journals.

THE ROYAL INSTITUTION

The Royal Institution of Great Britain was founded in 1799 by Benjamin Thompson, Count Rumford. It has occupied the same premises for nearly 200 years and, in that time, a truly astounding series of scientific discoveries has been made within its walls. Rumford himself was an early and effective exponent of energy conservation. Thomas Young established the wave theory of light; Humphry Davy isolated the first alkali and alkaline earth metals, and invented the miners' lamp; Tyndall explained the flow of glaciers and was the first to measure the absorption and radiation of heat by gases and vapours; Dewar liquefied hydrogen and gave the world the vacuum flask; all who wished to learn the new science of X-ray crystallography that W. H. Bragg and his son had discovered came to the Royal Institution, while W. L. Bragg, a generation later, promoted the application of the same science to the unravelling of the structure of proteins. In the recent past the research concentrated on photochemistry under the leadership of Professor Sir George (now Lord) Porter, while the current focus of the research work is the exploration of the properties of complex materials.

Towering over all else is the work of Michael Faraday, the London bookbinder who became one of the world's greatest scientists. Faraday's discovery of electromagnetic induction laid the foundation of today's electrical industries. His magnetic laboratory, where many of his most important discoveries were made, was restored in 1972 to the form it was known to have had in 1845. A newly created museum, adjacent to the laboratory, houses a unique collection of original apparatus arranged to illustrate the more important aspects of Faraday's immense contribution to the advancement of science in his fifty year's work at the Royal Institution.

Why the Royal Institution is Unique

It provides the only forum in London where non-specialists may meet the leading scientists of our time and hear their latest discoveries explained in everyday language.

It is the only Society that is actively engaged in research, and provides lectures covering all aspects of science and technology, with membership open to all.

It houses the only independent research laboratory in London's West End (and one of the few in Britain)—the Davy Faraday Research Laboratory.

What the Royal Institution Does for Young Scientists

The Royal Institution has an extensive programme of scientific activities designed to inform and inspire young people. This programme includes lectures for primary and secondary school children, sixth form conferences, Computational Science Seminars for sixth-formers and Mathematics Master classes for 12–13 year-old children.

What the Royal Institution Offers to its Members

Programmes, each term, of activities including summaries of the Discourses; synopses of the Christmas Lectures and annual Record.

Evening Discourses and an associated exhibition to which guests may be invited.

An annual volume of the *Proceedings of the Royal Institution of Great Britain* containing accounts of Discourses.

Christmas Lectures to which children may be introduced.

Meetings such as the RJ Discussion evenings; Seminars of the Royal Institution Centre for the History of Science and Technology, and other specialist research discussions.

Use of the Libraries and borrowing of the books. The Library is open from 9 a.m. to 9 p.m. on weekdays

Use of the Conversation Room for social purposes.

Access to the Faraday Laboratory and Museum for themselves and guests.

Invitations to debates on matters of current concern, evening parties and lectures marking special scientific occasions.

Royal Institution publications at privileged rates.

Group visits to various scientific, historical, and other institutions of interest.

Evening Discourses

The Evening Discourses have been given regularly since 1826. They cover all aspects of science and technology (with regular ventures into the arts) in a form suitable for the interested layman, and many scientists use them to keep in touch with fields other than their own. An exhibition, on a subject relating to the Discourse, is arranged each evening, and light refreshments are available after the lecture.

Christmas Lectures

Faraday introduced a series of six Christmas Lectures for children in 1826. These are still given annually, but today they reach a much wider audience through television. Titles have included: 'The Languages of Animals' by David Attenborough, 'The Natural History of a Sunbeam' by Sir George Porter, 'The Planets' by Carl Sagan and 'Exploring Music' by Charles Taylor.

The Library

The Library contains over 40,000 volumes, and is particularly strong in long runs of scientific periodicals. It has a fine collection of the history of science and science for the non-specialist. A selection of newspapers and magazines is available in the Conversation Room which is still characteristic of a library of the early nineteenth century.

Schools Lectures

Extending the policy of bringing science to children, the Royal Institution provides lectures throughout the year for school children of various ages, ranging from primary to sixth-form groups. These lectures, attended by thousands, play a vital part in stimulating an interest in science by means of demonstrations, many of which could not be performed in schools.

Seminars, Masterclasses, and Primary Schools Lectures

In addition to educational activities within the Royal Institution, there is an expanding external programme of activities which are organized at venues throughout the UK. These include a range of seminars and masterclasses in the areas of mathematics, technology and, most recently, computational science. Lectures aimed at the 8–10 year-old age group are also an increasing component of our external activities.

Teacher workshops

Lectures to younger children are commonly accompanied by workshops for teachers which aim to explain, illustrate, and amplify the scientific principles demonstrated by the lecture.

Membership of the Royal Institution
Member

The Royal Institution welcomes all who are interested in science, no special scientific qualification being required. By becoming a Member of the Royal Institution an individual not only derives a great deal of personal benefit and enjoyment but also the satisfaction of helping to support the unique contribution made to our society by the Royal Institution.

Family Associate Subscriber

A Member may nominate one member of his or her family residing at the same address, and not being under the age of 16 (there is no upper age limit), to be a Family Associate Subscriber. Family Associate Subscribers can attend the Evening Discourses and other lectures, and use the Libraries.

Associate Subscriber

Any person between the ages of 16 and 27 may be admitted as an Associate Subscriber. Associate Subscribers can attend the Evening Discourses and other lectures, and use the Libraries. (On transferring to full Membership, Associate Subscribers are not required to pay the usual admission fee.)

Junior Associate

Any person between the ages of 11 and 15 may be admitted as a Junior Associate. Junior Associates can attend the Christmas Lectures and other functions, and use the Libraries. Junior Associates may take part in educational visits organized during Easter and Summer vacations.

Corporate Subscriber

Companies, firms and other bodies are invited to support the work of the Royal Institution by becoming Corporate Subscribers; such organizations make a very valuable contribution to the income of the Institution and so endorse its value to the community. Two representatives may attend the Evening Discourses and other lectures, and may use the Libraries.

College Corporate Subscriber

Senior educational establishments may become College Corporate Subscribers; this entitles two representatives to attend the Evening Discourses and other lectures, and to use the libraries.

School Subscriber

Schools and Colleges of Education may become School Subscribers; this entitles two members of staff to attend the Evening Discourses and other lectures, and to use the Libraries.

Membership forms can be obtained from: The Membership Secretary, The Royal Institution, 21 Albermarle Street, London W1X 4BS.

Telephone: 071 409 2992. Fax: 071 629 3569